The Ultimate Math Survival Guide

- Part 2 -

Richard W. Fisher

The Ultimate Math Survival Guide, Part 2
ISBN 13: 978-0-9843629-6-7

Table of Contents

Geometry

Problem Solving

Pre-Algebra

Answer Key

Notes to the Teacher or Parent

What sets this book apart from other books is its approach. It is not just a math book, but a system of teaching math. Each daily lesson contains three key parts: **Review Exercises**, **Helpful Hints**, and **Problem Solving**. Teachers have flexibility in introducing new topics, but the book provides them with the necessary structure and guidance. The teacher can rest assured that essential math skills in this book are being systematically learned.

This easy-to-follow program requires only fifteen or twenty minutes of instruction per day. Each lesson is concise and self-contained. The daily exercises help students to not only master math skills, but also maintain and reinforce those skills through consistent review - something that is missing in most math programs. Skills learned in this book apply to all areas of the curriculum, and consistent review is built into each daily lesson. Teachers and parents will also be pleased to note that the lessons are quite easy to correct.

This book is based on a system of teaching that was developed by a math instructor over a thirty-year period. This system has produced dramatic results for students. The program quickly motivates students and creates confidence and excitement that leads naturally to success.

Please read the following "How to Use This Book" section and let this program help you to produce dramatic results with your math students.

How to Use This Book

This book is best used on a daily basis. The first lesson should be carefully gone over with students to introduce them to the program and familiarize them with the format. It is hoped that the program will help your students to develop an enthusiasm and passion for math that will stay with them throughout their education.

As you go through these lessons every day, you will soon begin to see growth in the student's confidence, enthusiasm, and skill level. The students will maintain their mastery through the daily review.

Step 1

The students are to complete the review exercises, showing all their work. After completing the problems, it is important for the teacher or parent to go over this section with the students to ensure understanding.

Step 2

Next comes the new material. Use the "Helpful Hints" section to help introduce the new material. Be sure to point out that it is often helpful to come back to this section as the students work independently. This section often has examples that are very helpful to the students.

Step 3

It is highly important for the teacher to work through the two sample problems with the students before they begin to work independently. Working these problems together will ensure that the students understand the topic, and prevent a lot of unnecessary frustration. The two sample problems will get the students off to a good start and will instill confidence as the students begin to work independently.

Step 4

Each lesson has problem solving as the last section of the page. It is recommended that the teacher go through this section, discussing key words and phrases, and also key strategies. Problem solving is neglected in many math programs, and just a little work each day can produce dramatic results.

Step 5

Solutions are located in the back of the book. Teachers may correct the exercises if they wish, or have the students correct the work themselves.

Other Titles Available

Mastering Essential Math Skills: Book 1/Grades 4-5

Mastering Essential Math Skills: Book 2/Middle Grades/High School

Whole Numbers and Integers

Fractions

Decimals and Percents

Geometry

Problem Solving

Pre-Algebra Concepts

No-Nonsense Algebra

The Secret to Success in Math

There are three clusters of math skills that every student needs to master. Students who learn and fully understand these essential topics can be considered algebra-ready. These skills are referred to as the Critical Foundations of Algebra.

Algebra-readiness is of huge importance. Algebra is the gateway subject to more advanced math, science, and technical classes. In turn, success in these classes will open a vast number of educational as well as career opportunities. In essence, success in math, and more specifically, Algebra, is a vital part of all students' education. Algebra-readiness will have a profound impact on success in school, college, career, and everyday life. Success in Algebra will open many doors for students. Unfortunately, those students who do not experience this success will find these same doors slammed shut.

Here are the Critical Foundations of Algebra:

- Whole Numbers—Students need to fully understand place value, and this must include a grasp of the meaning of the basic operations of addition, subtraction, multiplication, and division. They will also need the knowledge of how to apply the operations to problem solving. Instant recall of number facts is important. Whole number operations rest on the automatic recall of addition and related subtraction facts, and of multiplication and related division facts. These number facts are to math as the letters of the alphabet are to reading.

- Fractions—Students need to fully understand fractions, including decimals, and percents. This includes positive and negative fractions. They will need to be able to use all of these in problem solving. Fractions represent a major obstacle to a high percentage of students. Fractions, decimals, and percents need to be thoroughly understood.

- Some Aspects of Geometry and Measurements—Experience with similar triangles is directly relevant for the study of Algebra. Also, knowledge of slope of a line and linear functions is very important. Students should understand the properties of two and three-dimensional shapes and be able to determine perimeter area, volume and surface areas. They should also be able to find unknown lengths, angles, and areas. As with whole numbers and fractions, applying geometric skills to problem solving is essential.

For all these skills, conceptual understanding, computational fluency, and problem-solving skills are each essential.

The Critical Foundations of Algebra identified here are not meant to comprise a complete preschool-to-algebra curriculum. However, when these skills are mastered, success in Algebra will be assured.

The great news is that *The Ultimate Math Survival Guide: Part I* and *Part II* will ensure that students learn and master the Critical Foundations of Algebra!

The Ultimate Math Survival Guide Part I includes the following:

- Whole Numbers & Integers
- Fractions
- Decimals & Percents

The Ultimate Math Survival Guide Part II includes the following:

- Geometry
- Problem Solving
- Pre-Algebra

It should be noted that all of these skills are presented in a simple, easy-to-understand format. It is my belief, that ALL students can be successful in math! All that they need is the proper guidance.

Sincerely,
Richard W. Fisher
Author

Section 1

Geometry

Review Exercises

Note to students and teachers: This section will include daily review from all topics covered in this book. Here are some simple problems with which to get started.

1. $\begin{array}{r} 345 \\ 16 \\ +\ 724 \\ \hline \end{array}$

2. $\begin{array}{r} 715 \\ -\ 79 \\ \hline \end{array}$

3. $\begin{array}{r} 247 \\ \times\ 6 \\ \hline \end{array}$

4. $96 + 72 + 16 =$

5. $800 - 216 =$

6. $8 \times 394 =$

Helpful Hints

Geometric Term:	Point	Line	Plane	Line Segment	Ray
Example:	• P	A B (line)	•A •B •C (plane)	A B (segment)	A B (ray)
Symbol:	P	$\overleftrightarrow{AB}$	plane ABC	$\overline{AB}$	$\overrightarrow{AB}$

Use the figure to answer the following:

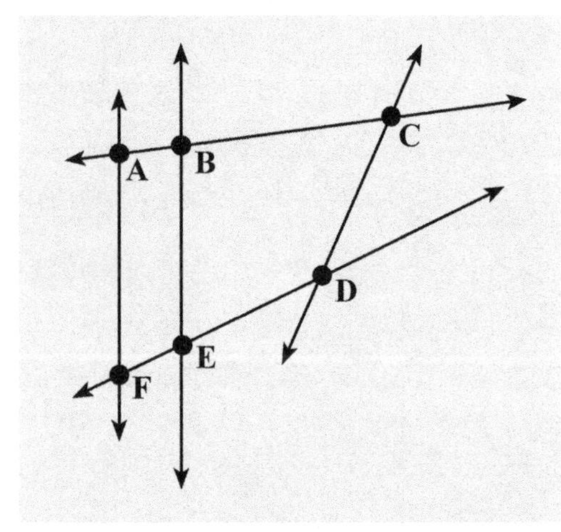

S1. Name 4 points

S2. Name 5 line segments

1. Name 5 lines

2. Name 5 rays

3. Name 3 points on $\overleftrightarrow{FD}$

4. Give another name for $\overleftrightarrow{AB}$

5. Give another name for $\overleftrightarrow{ED}$

6. Give another name for $\overleftrightarrow{AC}$

7. Name 2 line segments on $\overleftrightarrow{FD}$

8. Name 2 rays on $\overleftrightarrow{FE}$

9. Name 2 rays on $\overleftrightarrow{AC}$

10. What point is common to lines $\overleftrightarrow{FD}$ and $\overleftrightarrow{BE}$?

1.
2.
3.
4.
5.
6.
7.
8.
9.
10.
Score

Problem Solving

Ken earned 2,500 dollars in March and 3,752 dollars in April. What were his total earnings for the two months?

Review Exercises

1. 724 + 16 + 347 =

2. 506 − 397 =

3. 7 × 2,137 =

4.
```
   46
 × 23
```

5.
```
   753
    66
   124
 + 237
```

6.
```
  5,000
 −  787
```

Helpful Hints

Use what you have learned to answer the following questions.

Use the figure to answer the following:

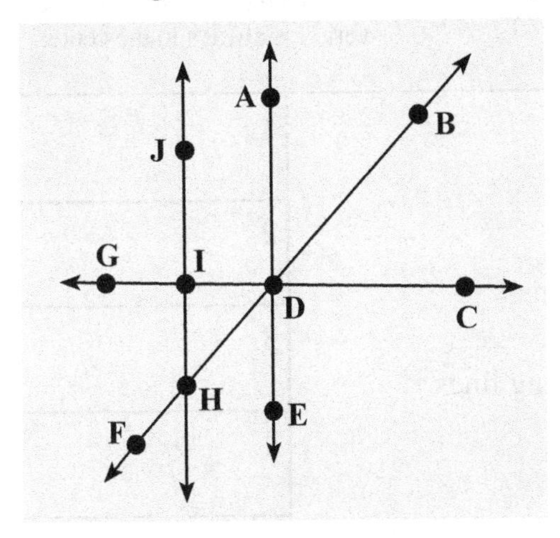

S1. Name 4 points

S2. Name 5 lines

1. Name 5 line segments

2. Name 5 rays

3. Name 3 points on $\overleftrightarrow{GC}$

4. Give another name for $\overleftrightarrow{HD}$

5. Give another name for $\overleftrightarrow{JI}$

6. Give another name for $\overrightarrow{ED}$

7. Name 2 rays on $\overleftrightarrow{FB}$

8. Name 2 line segments on $\overleftrightarrow{GC}$

9. Name 2 rays on $\overleftrightarrow{GC}$

10. What point is common to $\overleftrightarrow{ID}$ and $\overleftrightarrow{AE}$?

1.
2.
3.
4.
5.
6.
7.
8.
9.
10.
Score

Problem Solving

A factory can produce 350 cars per week. How many cars can the factory produce in one year? (Hint: How many weeks are there in a year?)

Review Exercises

1. $3\overline{)636}$
2. $5\overline{)617}$
3. $8\overline{)2,372}$
4. $7 \times 658 =$
5. $926 + 75 + 396 =$
6. $7,001 - 2,658 =$

Helpful Hints

Geometric Term:	Parallel Lines	Intersecting Lines	Perpendicular Lines	Angle
Example:	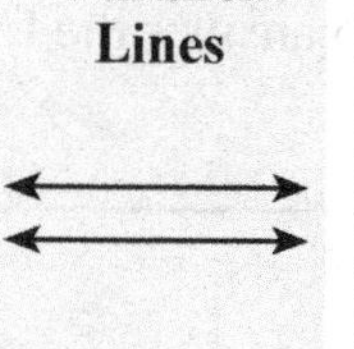	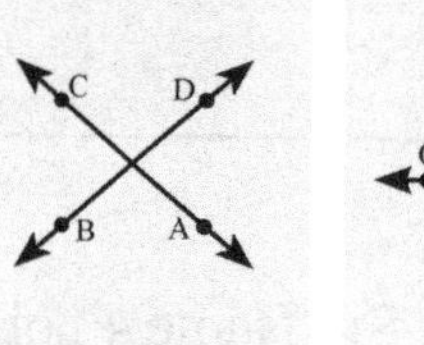	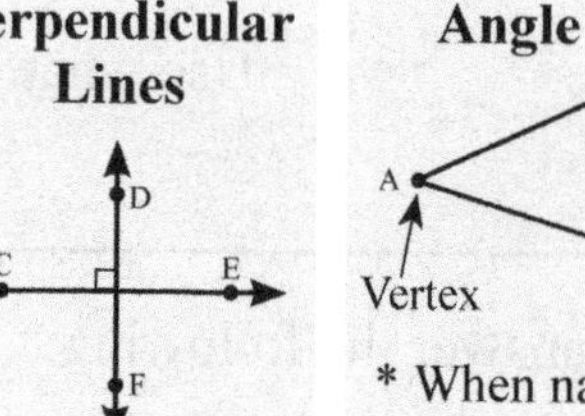	A, D, C, Vertex. Symbols: $\angle$ DAC, $\angle$ CAD, $\angle$ A * When naming an angle, the vertex is always in the center.

Use the figure to answer the following:

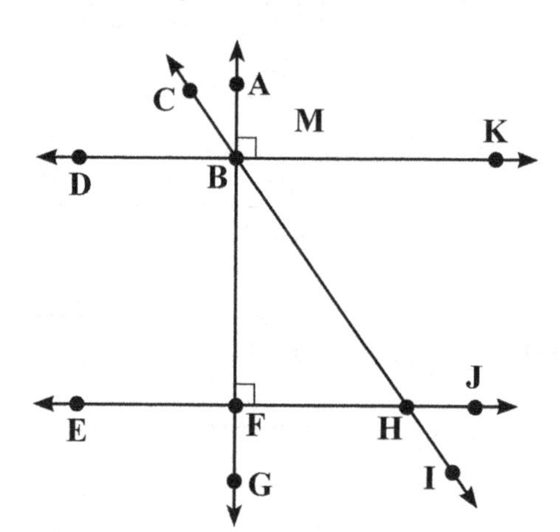

S1. Name 2 parallel lines

S2. Name 2 perpendicular lines

1. Name 3 pairs of intersecting lines
2. Name 5 angles
3. Name 3 angles that have **B** as their vertex.
4. Name 3 angles that have **H** as their vertex.
5. Name 3 lines
6. Name 5 line segments
7. Name 5 rays
8. Name 3 line segments on $\overleftrightarrow{\mathbf{BH}}$
9. Name 3 lines which include point **B**.
10. Give two other names for $\angle$**JHI**

1.
2.
3.
4.
5.
6.
7.
8.
9.
10.
Score

Problem Solving

Julio had test scores of 75, 96, 83, and 94.
What was his average score?

Review Exercises

1. Sketch two parallel lines.

2. Sketch an angle and label it ∠ABC.

3. Sketch 2 lines $\overleftrightarrow{AB}$ and $\overleftrightarrow{CD}$ that are perpendicular.

4. $\begin{array}{r} 906 \\ \times \ 8 \\ \hline \end{array}$

5. $\begin{array}{r} 7{,}112 \\ - \ 667 \\ \hline \end{array}$

6. $7\overline{)847}$

Helpful Hints

* When identifying an angle, the vertex is always in the center.

Example: In ∠CKD, K is the vertex.

Use the figure to answer the following:

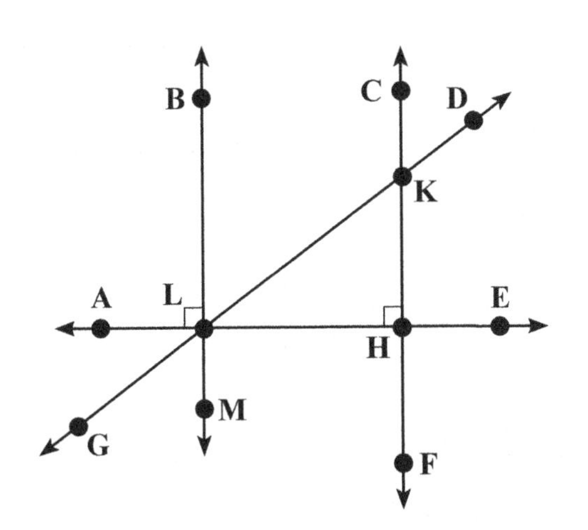

S1. Name 2 perpendicular lines.

S2. Name 2 parallel lines.

1. Name 4 angles.

2. Name 3 pairs of intersecting lines.

3. Name 2 angles with **L** as a vertex.

4. Name 2 angles with **K** as a vertex.

5. Name 4 lines.

6. Name 4 rays.

7. Name 4 line segments.

8. Name 3 lines which include point **H**.

9. Name 3 line segments on $\overleftrightarrow{\mathbf{AE}}$.

10. Give another name for angle ∠**BLH**.

1.
2.
3.
4.
5.
6.
7.
8.
9.
10.
Score

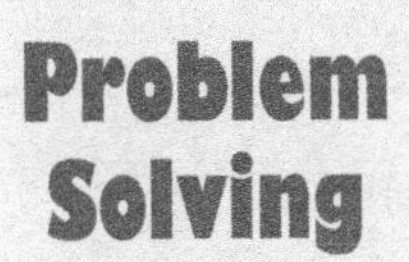

Six students earned $924. If they wanted to divide the money equally, how much would each of them receive?

Review Exercises

1. $852 + 276 + 19 =$

2. Sketch 2 lines $\overleftrightarrow{AB}$ and $\overleftrightarrow{CD}$ that are parallel.

3. $800 - 65 =$

4. Sketch 2 lines $\overleftrightarrow{MK}$ and $\overleftrightarrow{CD}$ that are intersecting.

5. 65×52

6. 246×70

Helpful Hints

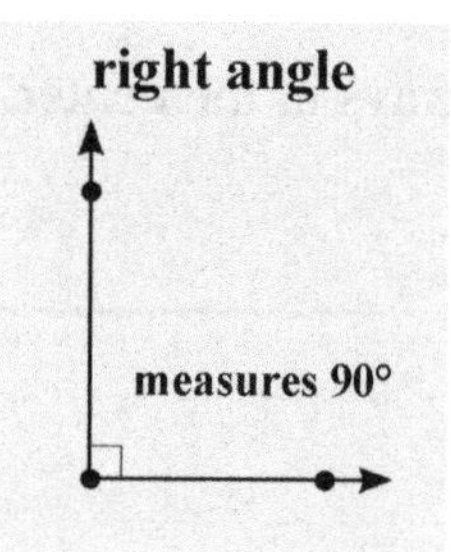

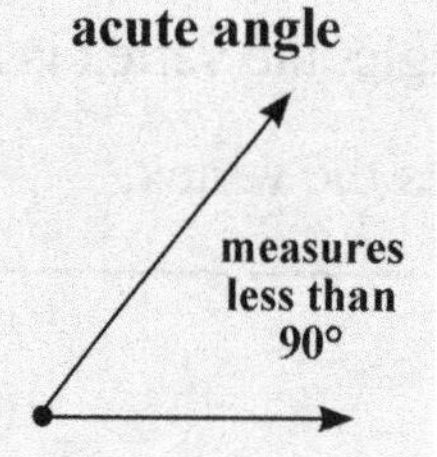

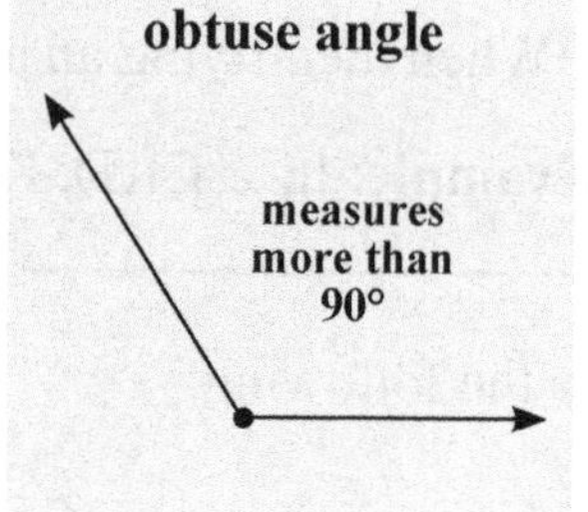

Use the figure to answer the following:

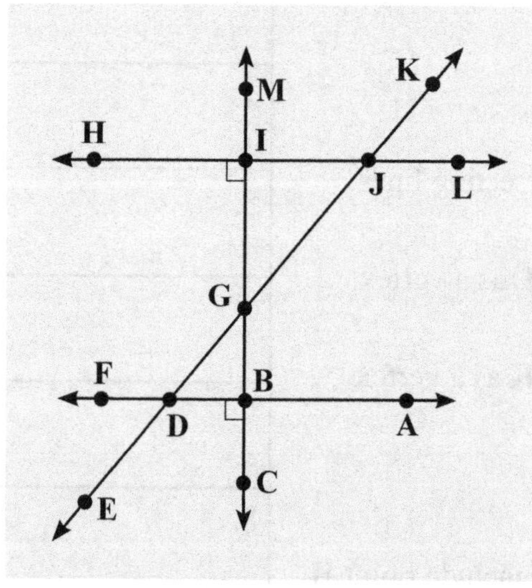

S1. Name 4 right angles.

S2. Name 5 acute angles.

1. Name 5 obtuse angles.
2. Name 5 straight angles.
3. What kind of angle is ∠**IJG**?
4. What kind of angle is ∠**EDB**?
5. What kind of angle is ∠**GBD**?
6. What kind of angle is ∠**GJK**?
7. Name an acute angle which has **J** as its vertex.
8. Name an obtuse angle which has **D** as its vertex.
9. Name a right angle which has **B** as its vertex.
10. Name a straight angle which has **D** as its vertex.

1.
2.
3.
4.
5.
6.
7.
8.
9.
10.
Score

Problem Solving

If a car traveled 275 miles in 5 hours, what was its average speed per hour?

Review Exercises

1. $77 + 888 + 666 =$

2. $5{,}012 - 763 =$

3. $6 \times 108 =$

4. 600×32

5. 365×402

6. $8\overline{)8{,}008}$

Helpful Hints

right angle

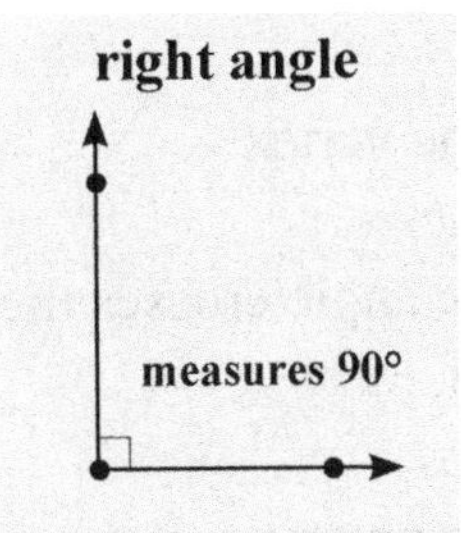

acute angle

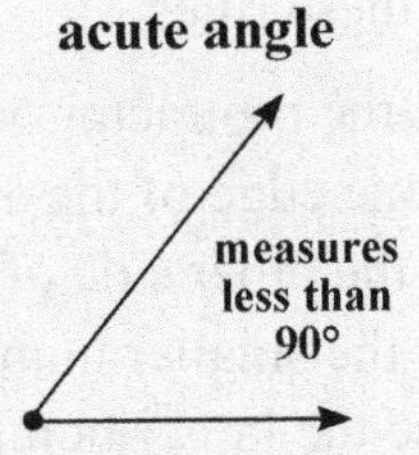

obtuse angle

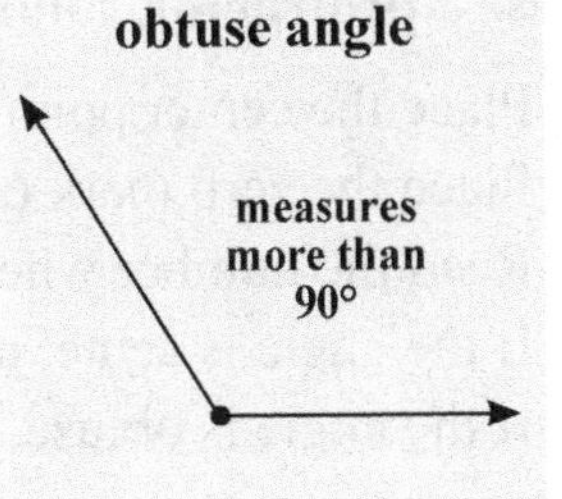

straight angle

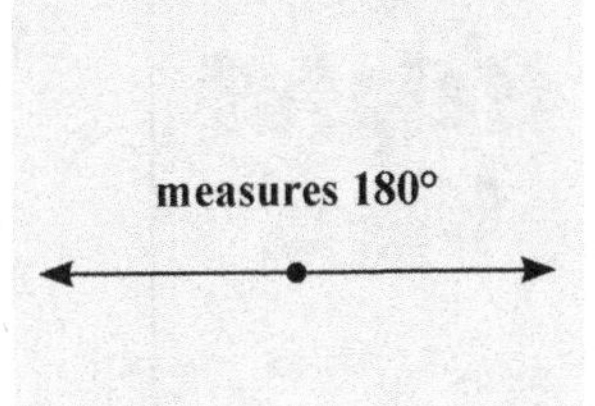

Use the figure to answer the following:

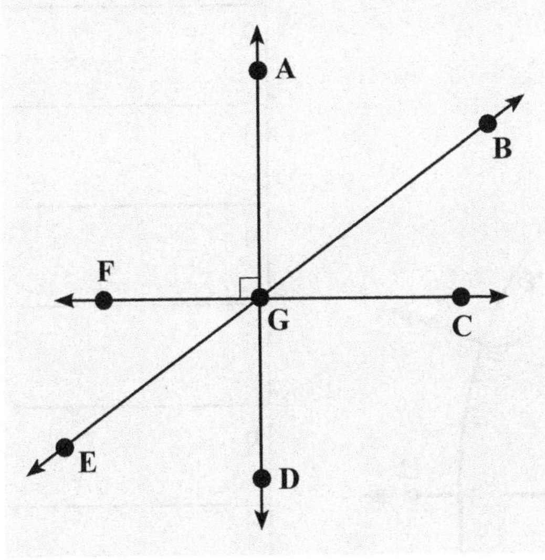

S1. Name 3 obtuse angles.

S2. Name 3 acute angles.

1. Name 2 straight angles.
2. Name 2 right angles.
3. What kind of angle is ∠BGC?
4. What kind of angle is ∠FGB?
5. What kind of angle is ∠DGF?
6. If ∠BGC is 30°, what is the measure of ∠AGB?
7. What kind of angle has a measure of 180°?
8. What kind of angle has a measure between 0° and 90°?
9. What kind of angle has a measure of between 90° and 180°.
10. If ∠FGE measures 30°, what is the measure of ∠EGC?

1.
2.
3.
4.
5.
6.
7.
8.
9.
10.
Score

Problem Solving

There are 12 rows of seats in a theater. Each row has 15 seats. If 153 seats are taken, how many are empty?

Review Exercises

1. What kind of angle is $\angle$**BAC**?

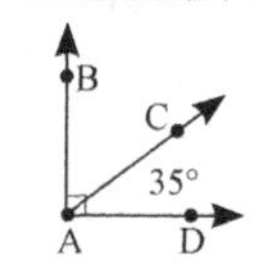

2. What kind of angle is $\angle$**ABC**?

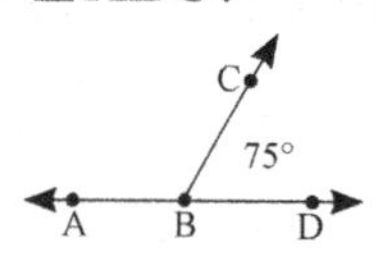

3. Sketch an acute angle named $\angle$**BDE**.

4. Sketch a right angle named $\angle$**MPQ**.

5. Sketch an obtuse angle named $\angle$**PTR**.

6. 9,001 − 767

Helpful Hints

To use a **protractor**, follow these rules.

1. Place the center point of the protractor on the vertex.
2. Place the zero mark on one edge of the angle.
3. Read the number where the other side of the angle crosses the protractor.
4. If the angle is acute, use the smaller number.
   If the angle is obtuse, use the larger number.

Use the figure to answer the questions. Classify the angle as right, acute, obtuse, or straight. Then tell how many degrees the angle measures.

S1. $\angle$**HBG** S2. $\angle$**DBH** 1. $\angle$**EBH** 2. $\angle$**CBH**

3. $\angle$**GBH** 4. $\angle$**DBA** 5. $\angle$**ABF** 6. $\angle$**FBH**

7. $\angle$**ABH** 8. $\angle$**ABG** 9. $\angle$**EBA** 10. $\angle$**FBA**

Problem Solving

420 students are placed into twenty equally sized classes. How many are there in each class?

1.
2.
3.
4.
5.
6.
7.
8.
9.
10.

Score

Review Exercises

1. Sketch an angle with a measure of 45°.

2. Sketch an angle with a measure of 110°.

3. Sketch lines $\overleftrightarrow{AC}$ and $\overleftrightarrow{DE}$ that are parallel.

4. Sketch lines $\overleftrightarrow{PQ}$ and $\overleftrightarrow{RM}$ that are perpendicular.

5.
$$\begin{array}{r} 976 \\ 42 \\ +\ 673 \\ \hline \end{array}$$

6.
$$\begin{array}{r} 5{,}101 \\ -\ 673 \\ \hline \end{array}$$

Helpful Hints

Use what you have learned to answer the following questions. Refer to the figure below.

Examples: **∠AMB = 180° - 145° = 35°**

∠DME = 85° - 20° = 65°

∠BMD = 145° - 85° = 60°

* Sometimes subtraction is necessary

Use the figure to find the measure of each angle. Next, classify the angle.

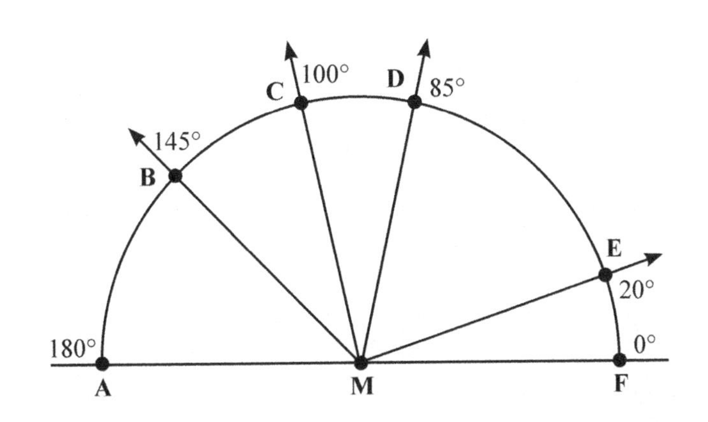

S1. ∠FME S2. ∠EMD 1. ∠FMC 2. ∠FMB

3. ∠BMC 4. ∠BMA 5. ∠FMA 6. ∠CME

7. ∠DMB 8. ∠EMA 9. ∠DMA 10. ∠EMB

1.
2.
3.
4.
5.
6.
7.
8.
9.
10.
Score

Problem Solving

A bakery produced 2,175 cookies. If they are put into boxes that contain 25 cookies, how many boxes are needed?

Review Exercises

1. $3\overline{)615}$
2. $9\overline{)1{,}989}$
3. 165×7
3. 308×72
5. $2{,}716 - 637$
6. $715 + 324 + 79 =$

Helpful Hints

When using a **protractor**, remember to follow these tips.

1. Place the center point of the protractor on the vertex of the angle.
2. Place the zero edge on one edge of the angle.
3. Read the number where the other side of the angle crosses the protractor.
4. If the angle is acute, use the smaller number.
   If the angle is obtuse, use the larger number.

With a protractor, measure the indicated angle in the figure. Tell the number of degrees. Also, classify the angle as acute, right, obtuse, or straight.

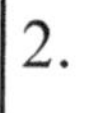

S1. ∠AMC	S2. ∠EMJ	1. ∠DMA	2. ∠FMJ
3. ∠FMA	4. ∠DMJ	5. ∠EMA	6. ∠CMJ
7. ∠IMA	8. ∠HMJ	9. ∠JME	10. ∠IMJ

1.
2.
3.
4.
5.
6.
7.
8.
9.
10.

Problem Solving

A car can travel 34 miles per gallon of gas. How many gallons will be consumed in traveling 408 gallons?

Score

Review Exercises

1. Sketch an obtuse angle named $\angle$**ABC**.

2. Sketch a right angle named $\angle$**EFG**.

3. Draw an acute angle named $\angle$**LMN**.

4. Draw two parallel lines $\overleftrightarrow{\mathbf{RS}}$ and $\overleftrightarrow{\mathbf{GH}}$.

5. What is the measure of angle $\angle$**BAC**?

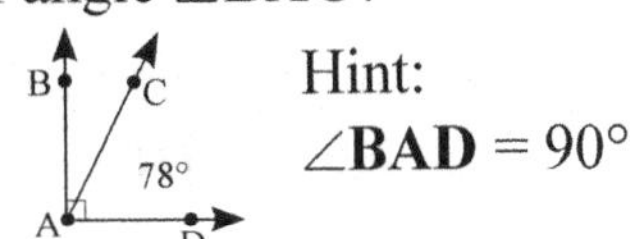

Hint: $\angle$**BAD** = 90°

6. What is the measure of angle $\angle$**TRW**?

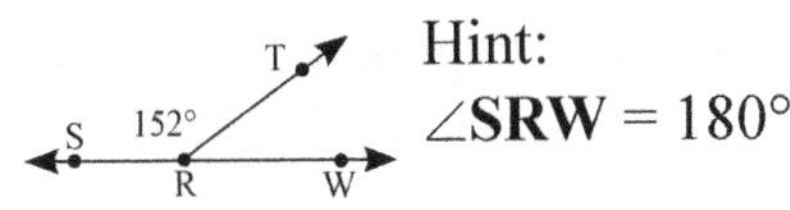

Hint: $\angle$**SRW** = 180°

Helpful Hints

Use what you have learned to answer the following questions.

With a protractor, measure the indicated angle in the figure. Tell the number of degrees. Also, classify the angle as acute, right, obtuse, or straight.

S1. $\angle$**LMJ** S2. $\angle$**AMG** 1. $\angle$**HML** 2. $\angle$**DML**

3. $\angle$**AMF** 4. $\angle$**AMC** 5. $\angle$**EML** 6. $\angle$**GMA**

7. $\angle$**GML** 8. $\angle$**BML** 9. $\angle$**AMD** 10. $\angle$**LMK**

1.
2.
3.
4.
5.
6.
7.
8.
9.
10.

Score

Problem Solving

A car traveled 32 miles using one gallon of gas.
How many gallons will be needed to travel 256 miles?
If the cost is $3.00 per gallon, how much will it cost?

Review Exercises

1. What is the measure of angle ∠**ABC**?

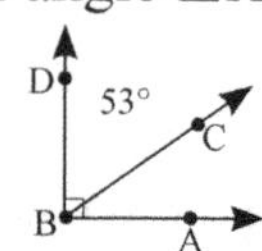

2. 968 × 7

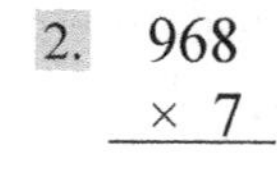

3. Sketch lines $\overleftrightarrow{AC}$ and $\overleftrightarrow{BD}$ that are intersecting.

4. Sketch an obtuse angle named ∠**MNO**.

5. Draw an acute angle named ∠**ADF**.

6. Draw a right angle named ∠**BCD**.

Helpful Hints

Two angles are **complementary** if the sum of their measures is 90°.
Two angles are **supplementary** if the sum of their measures is 180°.

Examples:

The complement of 30° is 60° because 30° + 60° = 90°

The supplement of 70° is 110° because 70° + 110° = 180°

Answer the following questions.

S1. Find the complement of 51°.

S2. Find the supplement of 87°.

For 1 through 5, find the complement of the given angle.

1. 72°
2. 16°
3. 5°
4. 52°
5. 71°

For 6 through 10, find the supplement of the given angle.

6. 75°
7. 12°
8. 15°
9. 172°
10. 113°

1.
2.
3.
4.
5.
6.
7.
8.
9.
10.
Score

Problem Solving

A man planned a three day, 72 mile hike. If he hiked 27 miles the first day, and 24 miles the second day, how many miles will he hike the third day?

Review Exercises

1. What is the supplement of 133° ?

2. What is the complement of 17° ?

3. 207×6

4. Draw two lines $\overleftrightarrow{AB}$ and $\overleftrightarrow{CD}$ that are parallel.

5. $736 + 75 + 937$

6. $7{,}112 - 3{,}143$

Helpful Hints

Use what you have learned to answer the following questions.

S1. Find the supplement of 65°.

S2. Find the compliment of 59°.

For 1 through 5, find the supplement of the given angle.

1. 79°
2. 141°
3. 33°
4. 19°
5. 62°

For 6 through 10, find the complement of the given angle.

6. 12°
7. 43°
8. 11°
9. 56°
10. 48°

1.
2.
3.
4.
5.
6.
7.
8.
9.
10.
Score

Problem Solving

Each package of paper contains 500 sheets. How many sheets of paper are there in 16 packages?

Review Exercises

1. What is the complement of 14° ?

2. What is the supplement of 14° ?

3. What is the measure of angle ∠**CDE**?

Hint: ∠**CDE** and ∠**EDF** are complementary angles.

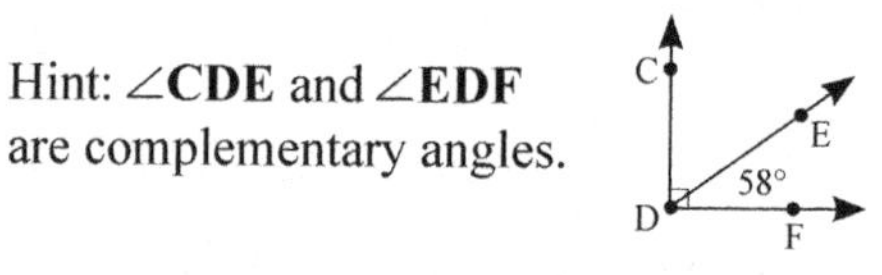

4. What is the measure of angle ∠**LMN?**

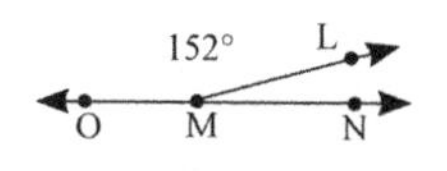

Hint: ∠**LMO** and ∠**LMN** are supplementary angles.

5. $6\overline{)1596}$

6.
$$\begin{array}{r} 7{,}123 \\ -\ \ 368 \\ \hline \end{array}$$

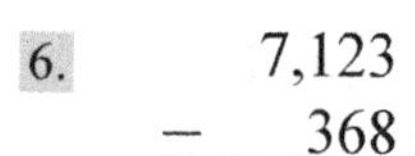

Helpful Hints

To draw an angle follow these steps.
1. Draw a ray and put the center point of the protractor on the end point.
2. Align the ray with the base line of the protractor.
3. Locate the degree of the angle you wish to draw.
4. Place a dot at that point and connect it to the endpoint of the ray.

Example: Draw an angle with measure 60°.

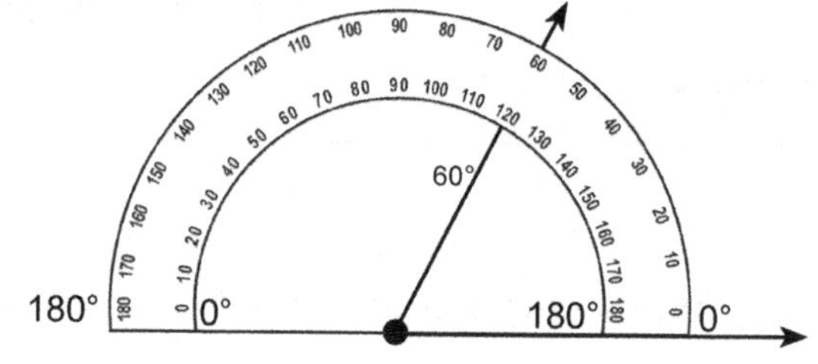

Use a protractor to draw angles with the following measures.

S1.	45°	S2.	120°	1.	30°
2.	110°	3.	70°	4.	52°
5.	90°	6.	160°	7.	40°

1.
2.
3.
4.
5.
6.
7.
8.
9.
10.
Score

Problem Solving

If Lola earns $4,500 per month, what is her annual income? (Hint: How many months are in a year?)

Review Exercises

1. What angle is supplementary to 82° ?

2. What angle is complementary to 17° ?

3. $\begin{array}{r} 367 \\ \times\ 15 \\ \hline \end{array}$

4. What is the measure of angle ∠**DEF**?

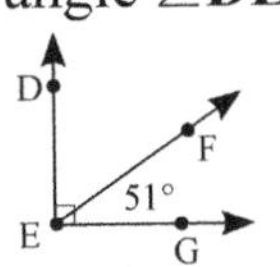

5. What is the measure of angle ∠**HIJ?**

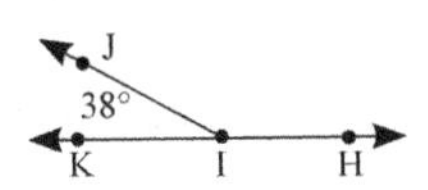

6. Draw two intersecting lines that intersect at point B.

Helpful Hints

Use what you have learned to draw the given angles.
*Be careful when lining up your protractor.

Use a protractor to draw angles with the following measures.

S1. 70°

S2. 150°

1. 85°

2. 20°

3. 55°

4. 125°

5. 35°

6. 95°

7. 65°

1.
2.
3.
4.
5.
6.
7.
8.
9.
10.
Score

Problem Solving

Adult tickets to a movie are nine dollars each and children's tickets are seven dollars each. How much would it cost for two adult tickets and four children's tickets?

Review Exercises

1. Name and classify.

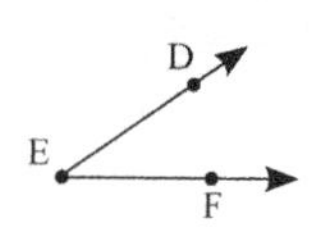

2. Name and classify.

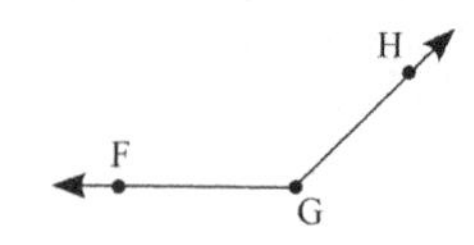

3. Name and classify.

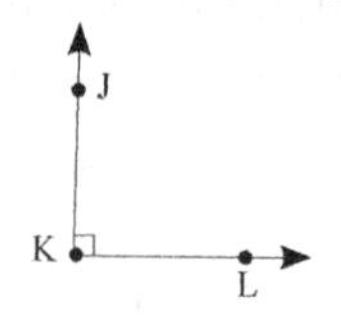

4. What kind of lines are these?

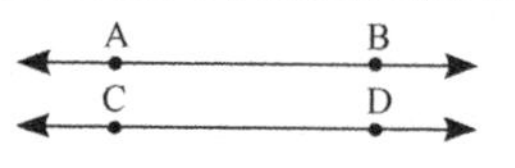

5. What kind of lines are these?

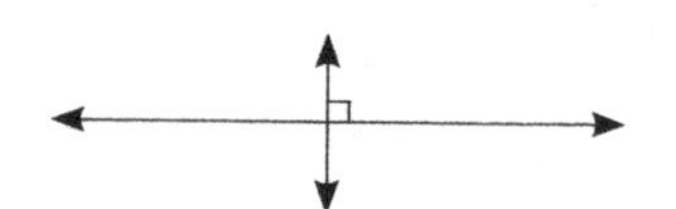

6. What is the complement of 15°

Helpful Hints

Polygons are closed figures made up of line segments.

Triangle — 3 sides

Rectangle — 4 sides, 4 right angles

Square — 4 congruent sides, 4 right angles

Parallelogram — 4 sides, opposite sides parallel

Trapezoid — 4 sides, 1 pair of parallel sides

Name each polygon. Some have more than one name.

S1. 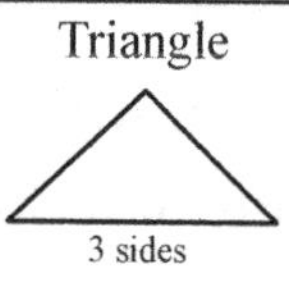

S2. 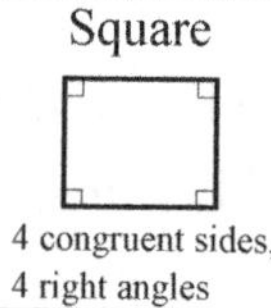

1. 

2. 

3. 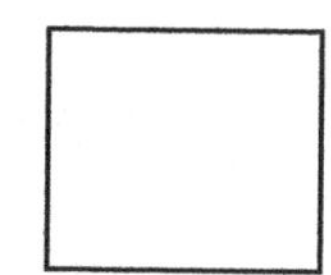

4. 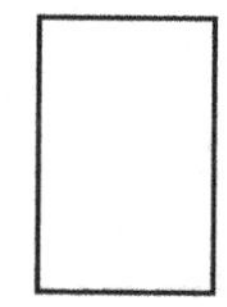

5. 

6. 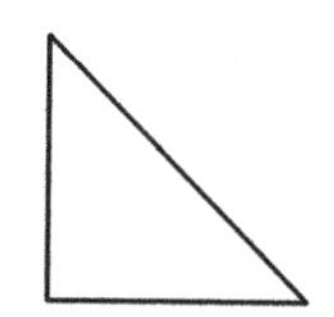

7. 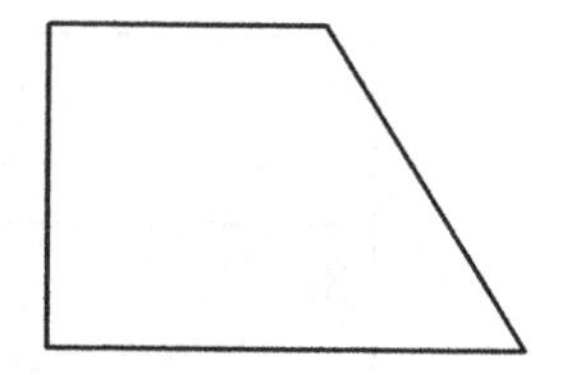

8. 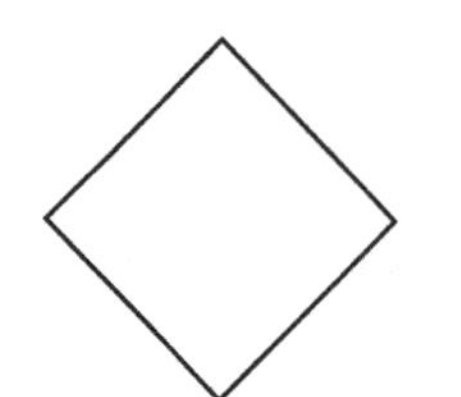

9. 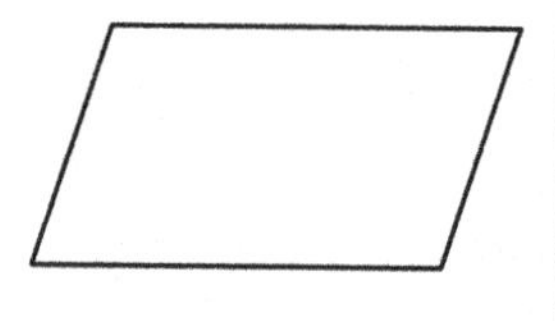

10. 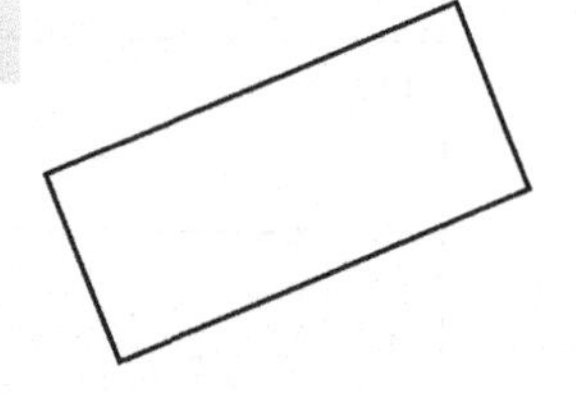

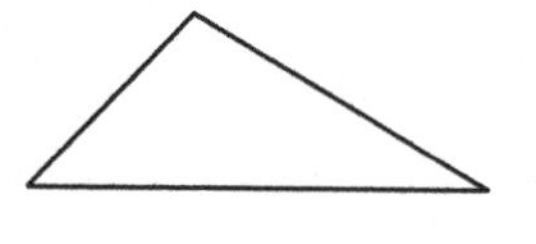

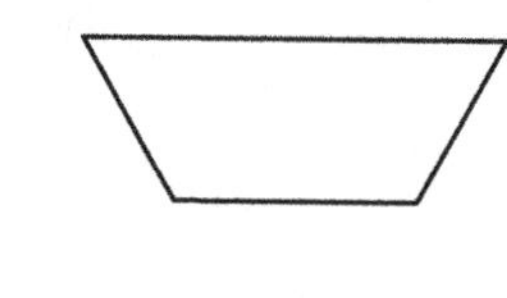

1.
2.
3.
4.
5.
6.
7.
8.
9.
10.

Score

Problem Solving

Claire needs to send out party invitations to 120 people. If cards come in boxes of 18, how many boxes does she need to buy? How many cards will be left over?

Review Exercises

1. What is the complement of 14°?

2. What are the two names for this figure?

5 in.
5 in. 5 in.
5 in.

3. What is the supplement of 112°?

4. Sketch an obtuse angle named ∠ **BDF**.

5. Sketch a right angle named ∠ **LMN.**

6. Sketch 2 perpendicular lines meeting at point B.

Helpful Hints

Use what you have learned to answer the following questions.
* Some have more than one name.

Name each polygon. Some have more than one name.

S1. 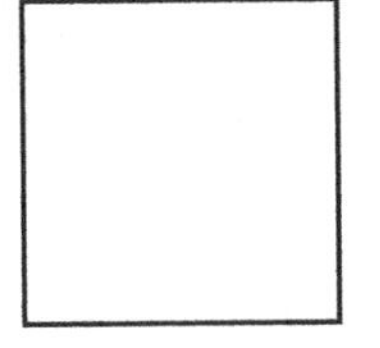

S2. 

1. 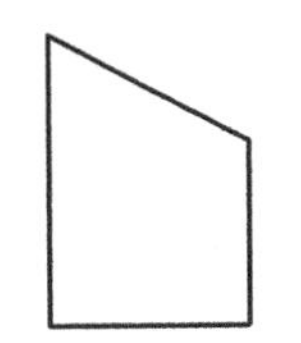

2. 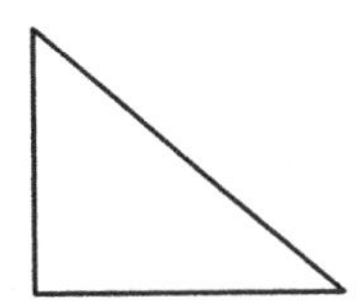

3. 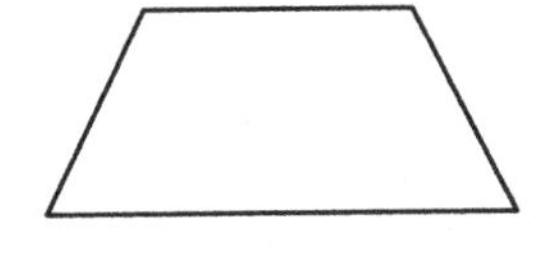

4. 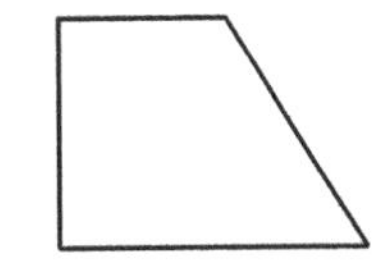

5. 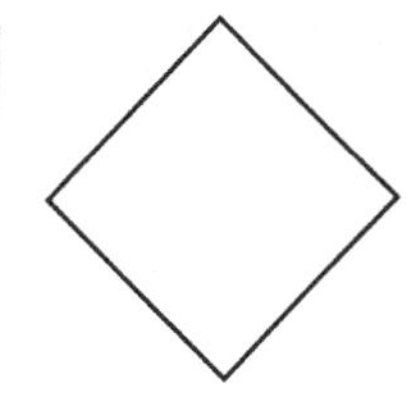

6. 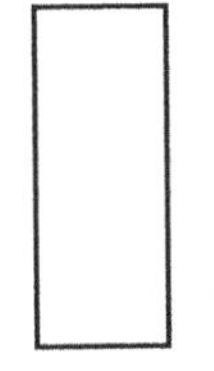

7. 

8. 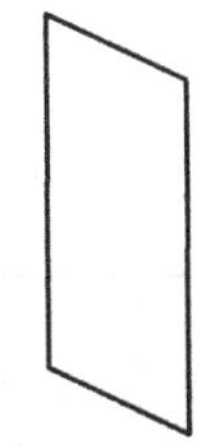

9. 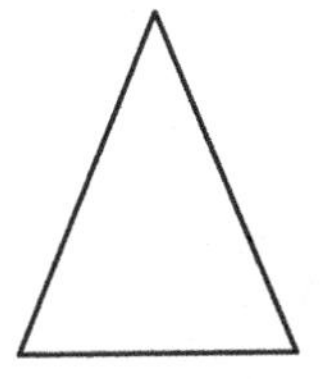

10. 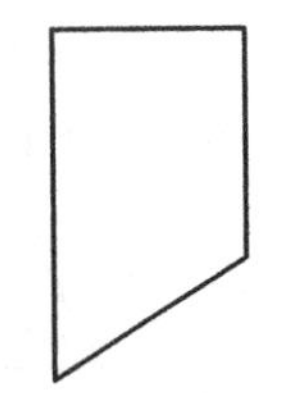

1.
2.
3.
4.
5.
6.
7.
8.
9.
10.
Score

Problem Solving

A school has 32 classrooms, each with 25 desks. If 743 students attend the school, how many desks will be left over?

Review Exercises

1. What polygon has three sides?

2. What polygon has one pair of parallel sides?

3. Name four polygons with four sides.

4. What polygons have 2 pairs of parallel sides?

5. What is the complement of 18°?

6. What is the supplement of 136°?

Helpful Hints	Triangles can be classified by sides and angles.	Sides			Angles		
		Equilateral	Scalene	Isosceles	Acute	Right	Obtuse
		3 congruent sides	No congruent sides	Two congruent sides	Three acute angles	One right angle	One obtuse angle

Classify each triangle by its sides and angles.

S1. Sides: ________ Angles: ________

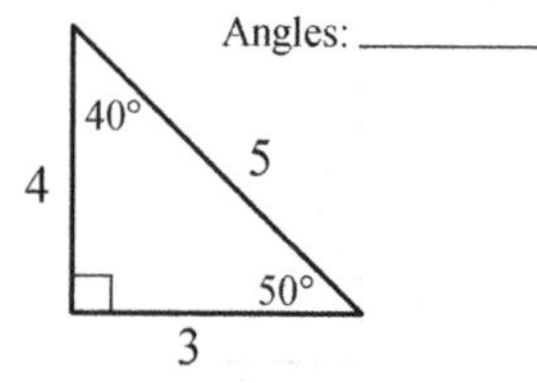

S2. Sides: ________ Angles: ________

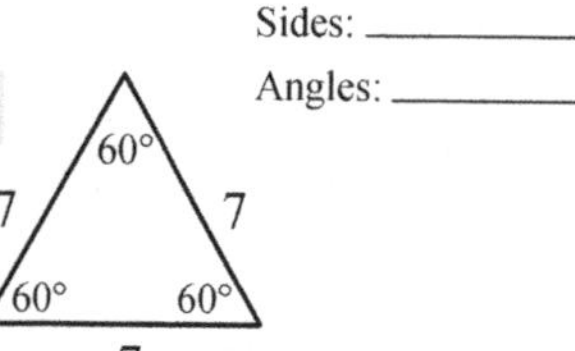

1. Sides: ________ Angles: ________

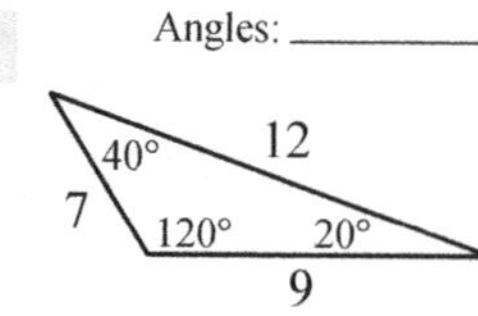

2. Sides: ________ Angles: ________

7, 80°, 7, 50°, 50°, 9

3. Sides: ________ Angles: ________

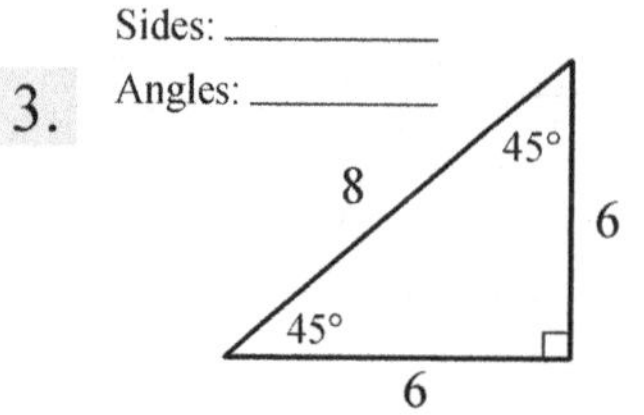

4. Sides: ________ Angles: ________

70°, 7, 3, 80°, 30°, 6

5. Sides: ________ Angles: ________

6. Sides: ________ Angles: ________

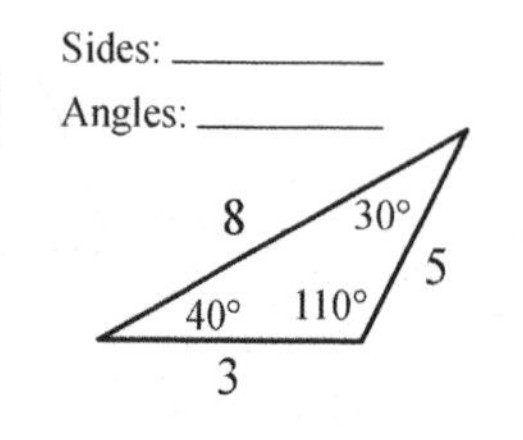

7. Sides: ________ Angles: ________

60°, 5, 3, 30°, 4

8. Sides: ________ Angles: ________

9. Sides: ________ Angles: ________

60°, 9 ft., 9 ft., 60°, 60°, 9 ft.

10. Sides: ________ Angles: ________

45°, 45°

1.
2.
3.
4.
5.
6.
7.
8.
9.
10.

Score

Problem Solving

Look at the triangles in the exercises above. When you add the three angles of any triangle, what will the total be?

Review Exercises

1. What kind of triangle has one obtuse angle?
2. What kind of triangle has three congruent sides?
3. What kind of triangle has all three sides of a different length?
4. What kind of triangle has two congruent sides?
5. If a triangle has three congruent sides what is the measure of each angle?
6. What kind of triangle has no congruent sides?

Helpful Hints

Use what you have learned to answer the following questions.

Classify each triangle by its side and angles using the facts given.
The facts are given in no particular order.

S1. Sides: ______
Angles: ______
Sides: 8, 8, 8
Angles: 60°, 60°, 60°

S2. Sides: ______
Angles: ______
Sides: 7, 8, 9
Angles: 70°, 60°, 50°

1. Sides: ______
Angles: ______
Sides: 7, 7, 10
Angles: 90°, 45°, 45°

2. Sides: ______
Angles: ______
Sides: 16, 10, 6
Angles: 30°, 110°, 40°

3. Sides: ______
Angles: ______
Sides: 12, 12, 12
Angles: 60°, 60°, 60°

4. Sides: ______
Angles: ______
Sides: 18, 14, 14
Angles: 80°, 50°, 50°

5. Sides: ______
Angles: ______
Sides: 3, 4, 5
Angles: 60°, 90°, 30°

6. Sides: ______
Angles: ______
Sides: 12, 16, 12
Angles: 45°, 45°, 90°

7. Sides: ______
Angles: ______
Sides: 9, 9, 9
Angles: 60°, 60°, 60°

8. Sides: ______
Angles: ______
Sides: 12, 6, 14
Angles: 30°, 70°, 80°

9. Sides: ______
Angles: ______
Sides: 6, 8, 8
Angles: 55°, 70°, 55°

10. Sides: ______
Angles: ______
Sides: 10, 8, 6
Angles: 40°, 90°, 50°

Answers
1.
2.
3.
4.
5.
6.
7.
8.
9.
10.
Score

Problem Solving

If a triangle has two angles with measures of 74° and 65° what is the measure of the third angle?

Review Exercises

1. Classify by sides.

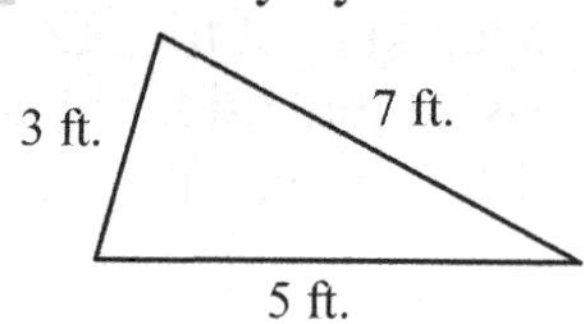

2. Classify by angles.

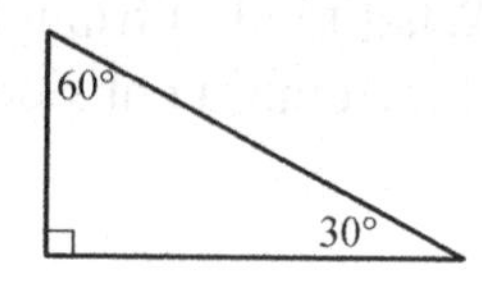

3. Classify by sides and angles.

Sides: ____________

Angles: ____________

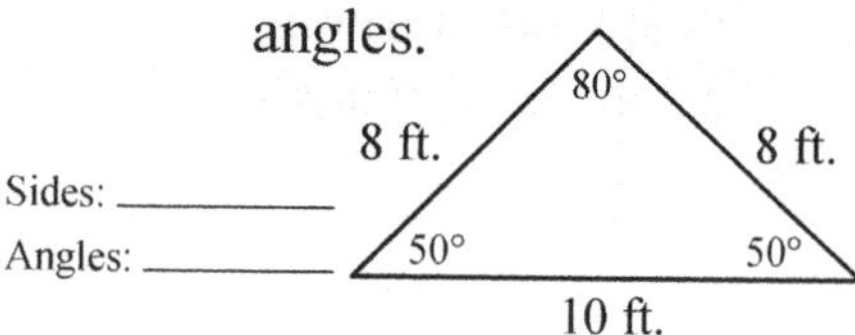

4. $\begin{array}{r} 756 \\ \times \quad 6 \\ \hline \end{array}$

5. $6\overline{)936}$

6. $500 - 334 =$

Helpful Hints

The distance around a polygon is its **perimeter**.

***A regular polygon has all sides congruent and all angles congruent.**

Examples:

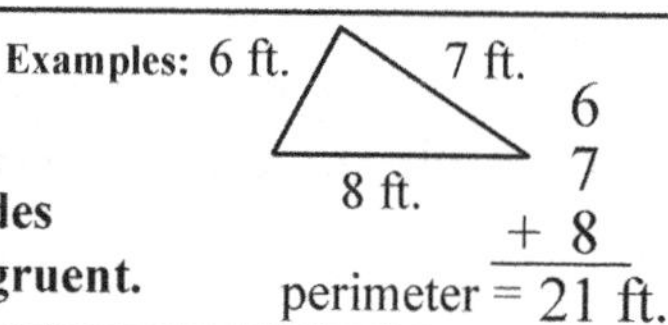

$\begin{array}{r} 6 \\ 7 \\ +\ 8 \\ \hline \end{array}$

perimeter = 21 ft.

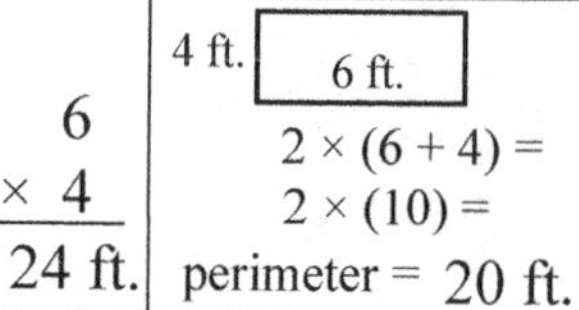

$\begin{array}{r} 6 \\ \times\ 4 \\ \hline \end{array}$

perimeter = 24 ft.

4 ft. [6 ft.]

$2 \times (6 + 4) =$

$2 \times (10) =$

perimeter = 20 ft.

Find the perimeter of each of the following.

S1.

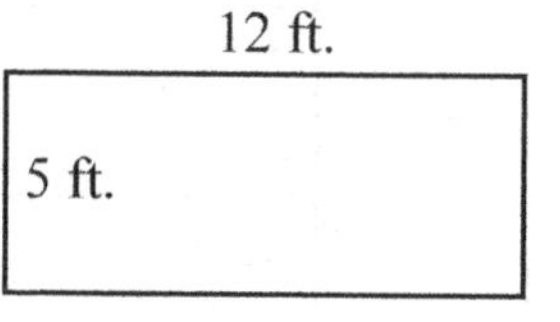

S2.

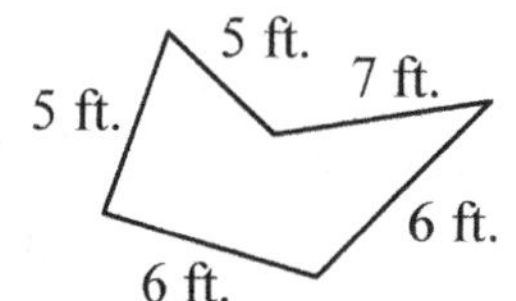

1.

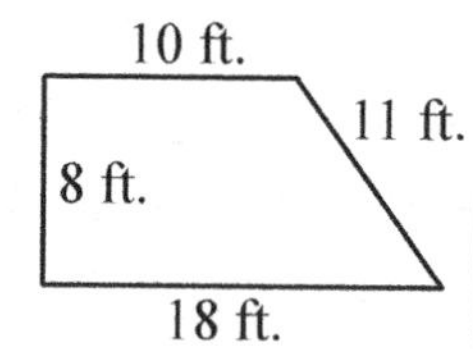

2.

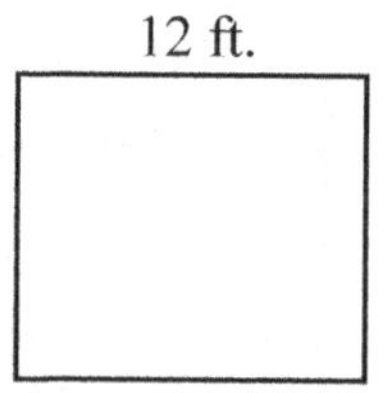

3.

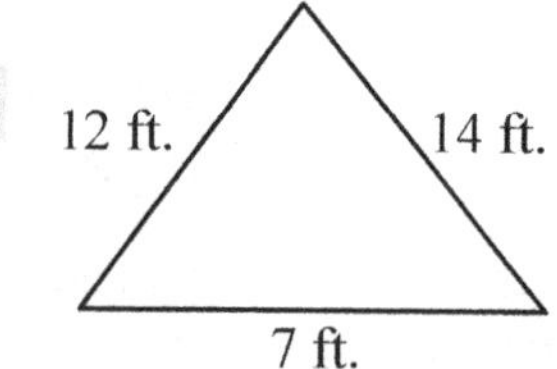

4.

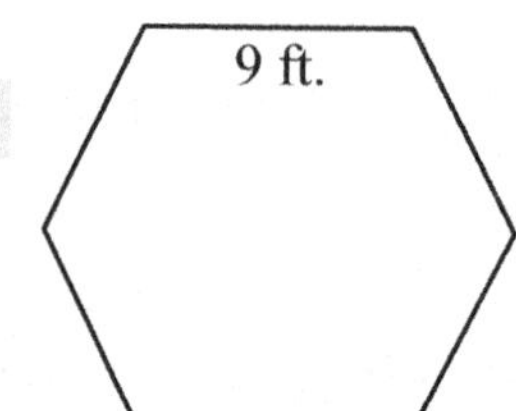

5.

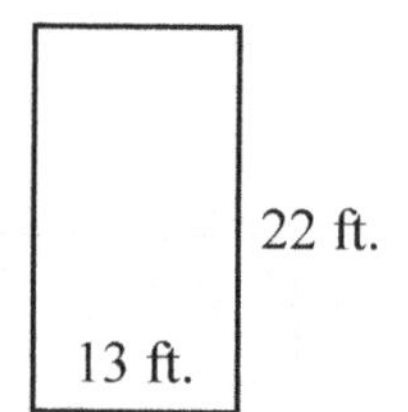

6.

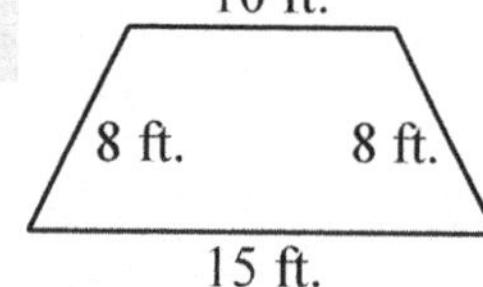

7.

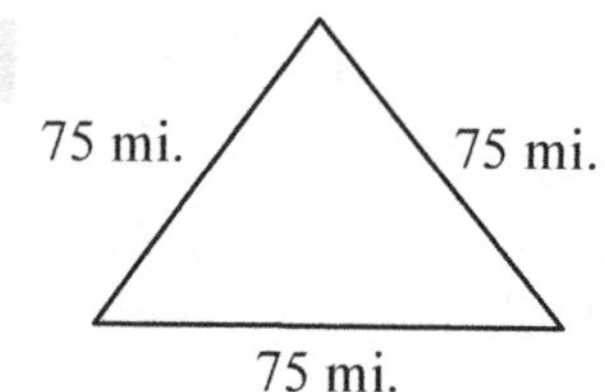

8.

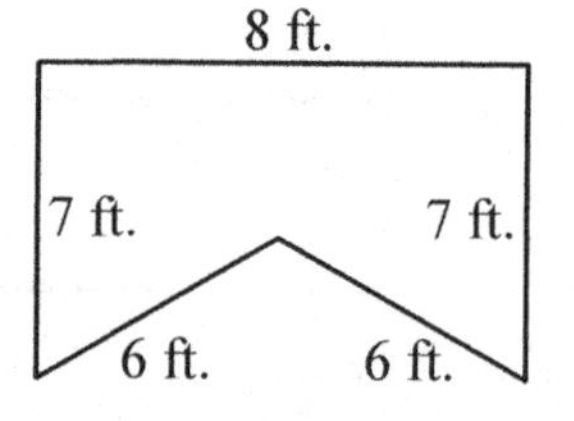

9.

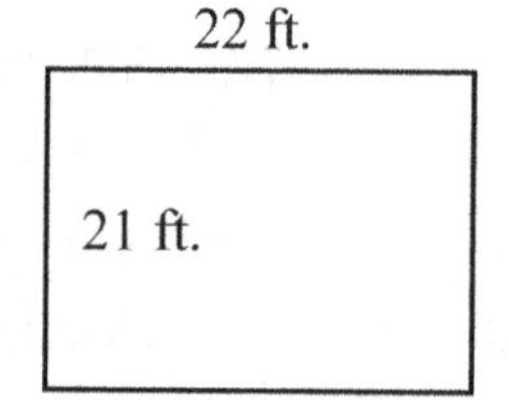

10.

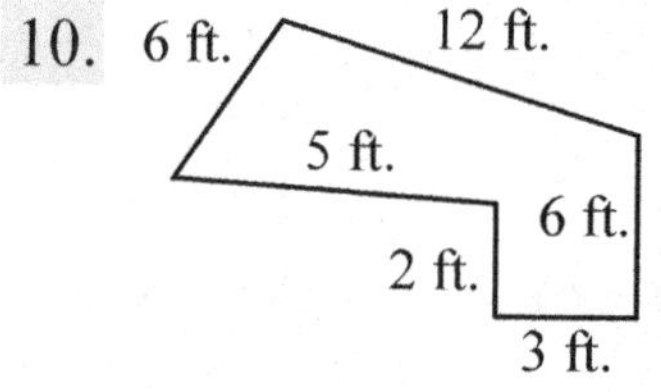

1. ____
2. ____
3. ____
4. ____
5. ____
6. ____
7. ____
8. ____
9. ____
10. ____

Score

Problem Solving

What is the perimeter of a rectangle with a length of 15 ft. and a width of 12 ft. (Draw a sketch before solving the problem.)

Review Exercises

1. Classify 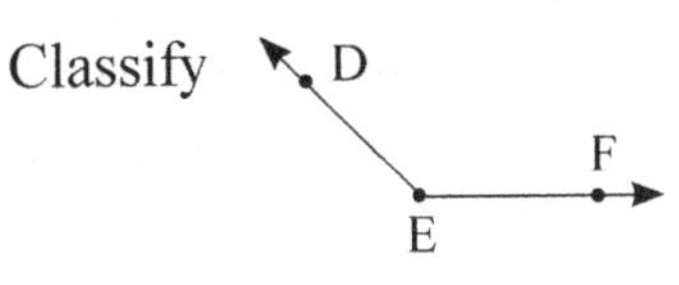

2. Classify 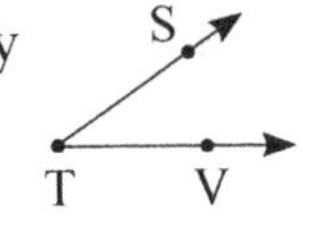

3. Classify 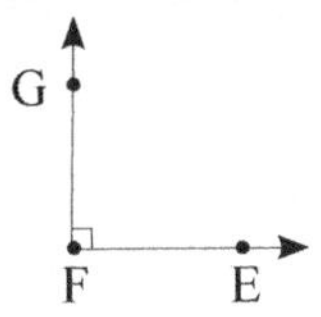

4. Classify by angles.

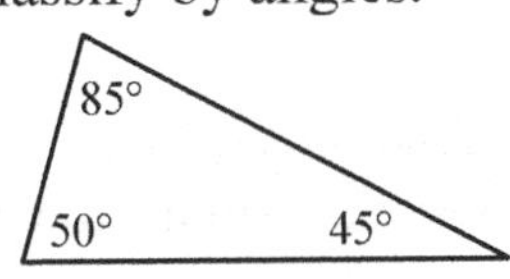

5. Classify by sides.

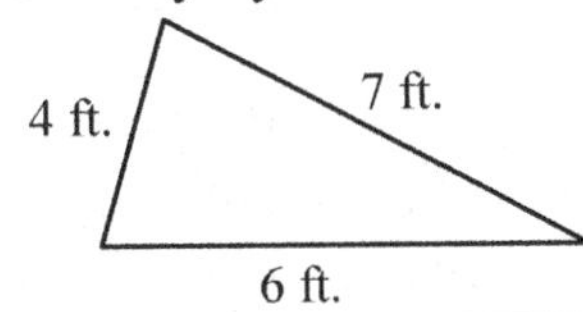

6. What is the supplement of 117°?

Helpful Hints

Use what you have learned to solve the following problems.

***Pentagon:** 5-sided polygon **Octagon:** 8-sided polygon
Hexagon: 6-sided polygon **Decagon:** 10-sided polygon

Find the perimeter of the following polygons.

S1. 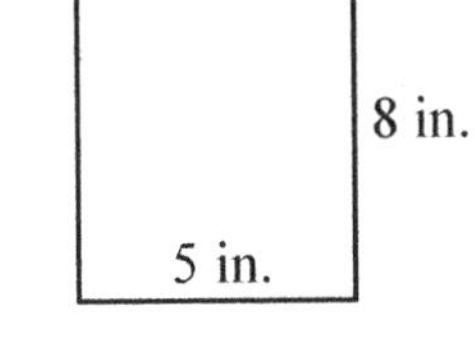

S2. 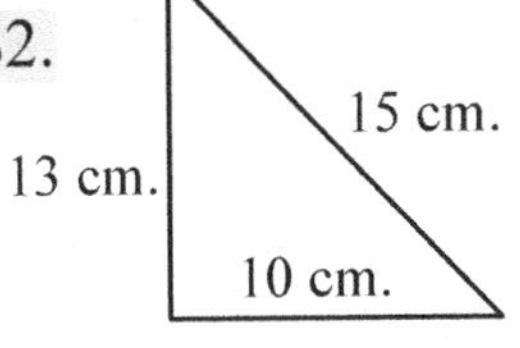

1. 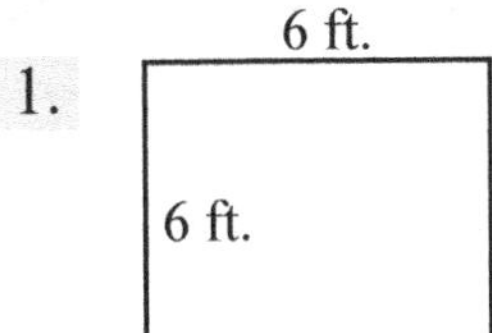

2. 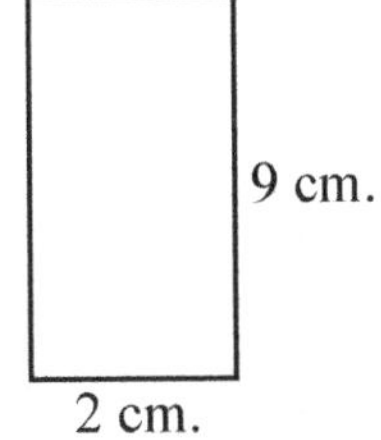

3. 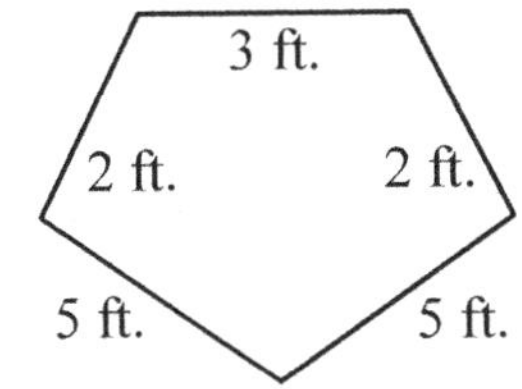

4. 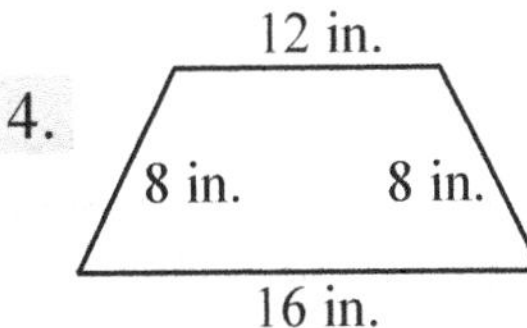

5. 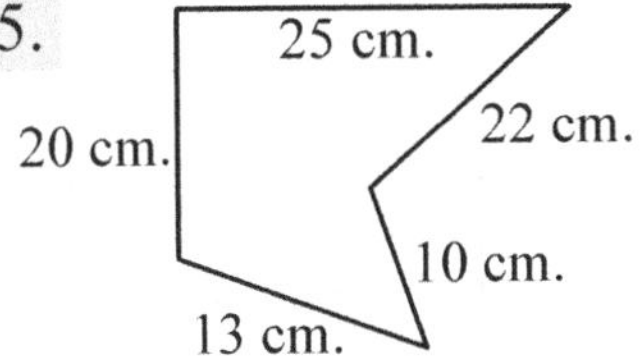

6. 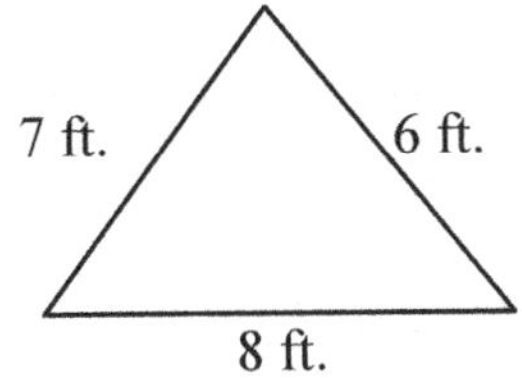

7. 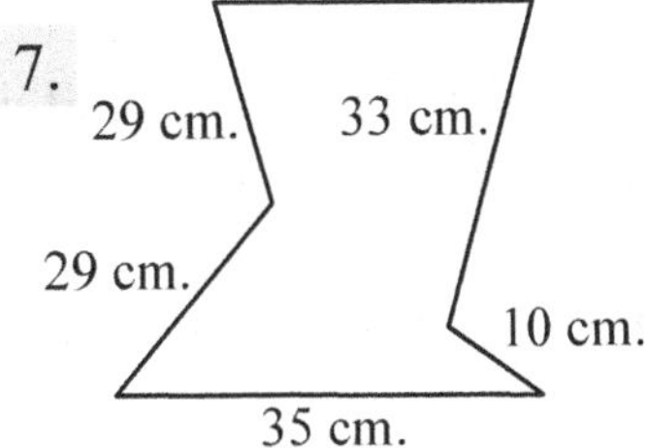

8. 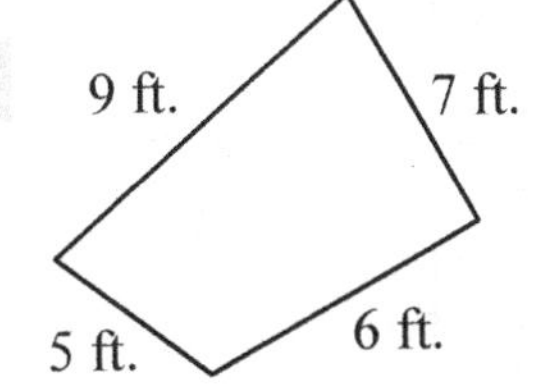

9. 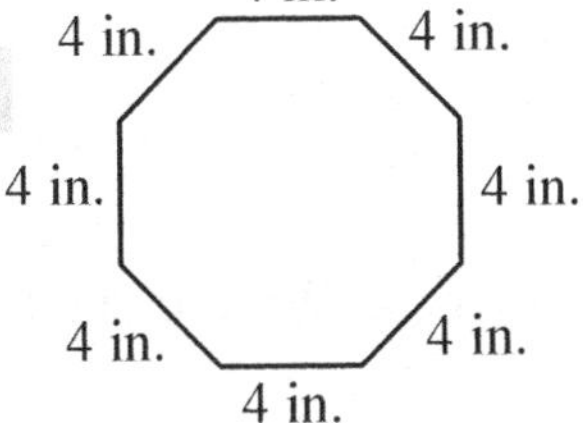

10. 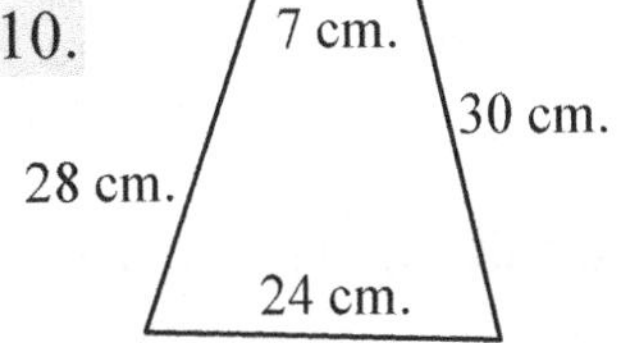

1.
2.
3.
4.
5.
6.
7.
8.
9.
10.
Score

Problem Solving

What is the perimeter of a regular hexagon with sides of 17 ft. each?

Review Exercises

1. Find the perimeter.

5 ft.

16 ft.

2. Find the perimeter.

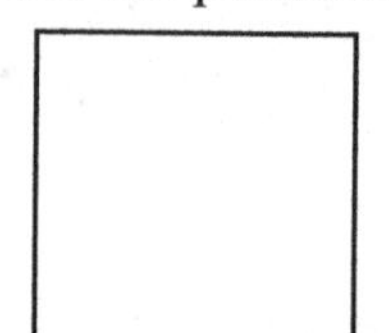

19 in.

3. Find the perimeter of a triangle with sides 17 ft., 20 ft., and 23 ft.

4. Find the perimeter of a regular octagon with sides of 12 ft.

5. What is the third angle of a triangle with angles of 75° and 65°?

6. What angle is supplementary to 63°?

Helpful Hints

When solving problems related to perimeter follow these directions.

1. Read the problem carefully to fully understand what is asked.
2. Draw a sketch.
3. Solve the problem.

S1. Jim wants to build a fence around his yard. It is in the shape of a rectangle with a length of 32 ft. and a width of 18 ft. how many feet of fencing material does he need to buy?

S2. Find the perimeter of a regular decagon that has sides of 52 ft.

1. What is the perimeter of a square with sides of 96 ft.?

2. Jolie wants to put a wood frame around a painting that is in the shape of a rectangle. If the length is 36 inches and the width is 18 inches, how many inches of wood frame will be needed?

3. A square has a perimeter of 156 ft. What is the length of each side?

4. A banner is in the shape of an equilateral triangle. If each side is 57 inches, what is the perimeter of the banner?

5. The perimeter of a regular hexagon is 138 inches. What is the length of each side?

6. Bill's yard is in the shape of a square with sides of 15 ft. If he wants to build a fence around the yard and materials are 12 dollars per foot, how much will the fence cost?

7. An equilateral triangle has a perimeter of 291 inches. What is the length of each side?

1.
2.
3.
4.
5.
6.
7.
Score

Problem Solving

Three pounds of steak costs $14.97. What is the cost per pound?

Review Exercises

For 1-6 draw a sketch and find the perimeter.

1. **Square:**
   Sides 32 ft.

2. **Rectangle:**
   Length 35 ft.
   Width 17 ft.

3. **Regular pentagon:**
   Sides 17 ft.

4. **Equilateral triangle:**
   Sides 113 inches

5. **Regular Hexagon:**
   Sides 17 inches

6. **Rectangle:**
   Length 116 ft.
   Width 90 ft.

Helpful Hints

Use what you have learned to solve the following problems.
*Remember to draw a sketch.

S1. A rectangular garden is 17 ft. by 12 ft. How many feet of fencing is needed to build a fence around the garden?

S2. A kitchen floor is in the shape of a square with sides of 18 ft. Mrs. Smith wants to trim the sides of the floor with wood. If wood trim comes in sections three feet long, how many sections must she buy?

1. How many feet is it around a square garden with sides of 17 ft.?

2. If a regular octagon has a perimeter of 128 inches, how long is each side?

3. A ranch is in the shape of a square with sides 10 miles long. If a runner can travel 5 miles per hour, how long will it take to run around the ranch?

4. The perimeter of a rectangle is 60 ft. If the length is 18 ft., what is the width?

5. Find the perimeter of a regular decagon with sides of 19 ft.

6. The perimeter of a square is 504 inches. What is the length of each side?

7. The sum of two sides of an equilateral triangle is 96 ft. What are the lengths of the sides of the triangle?

1.
2.
3.
4.
5.
6.
7.
Score

Problem Solving

A plane traveled 3,600 miles in eight hours.
What was its average speed per hour?

Review Exercises

1. Classify by sides

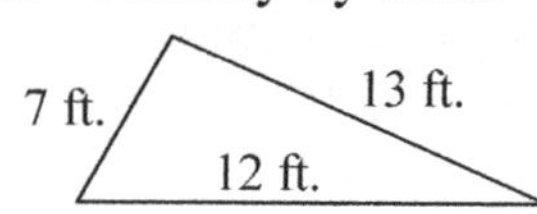

2. Classify by angles

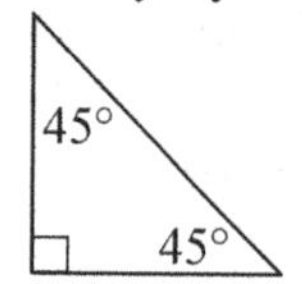

3. What angle is complementary to 13° ?

4. Find the perimeter of a square with sides 27 ft.

5. Two angles in a triangle are 90° and 30°. What is the measure of the third angle?

6. The perimeter of a square is 64 ft. What is the length of each side?

Helpful Hints

These are the parts of a **circle**.

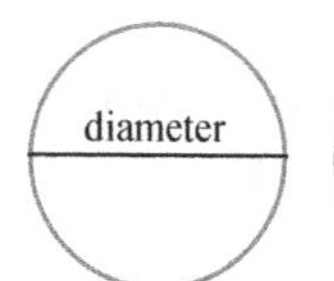

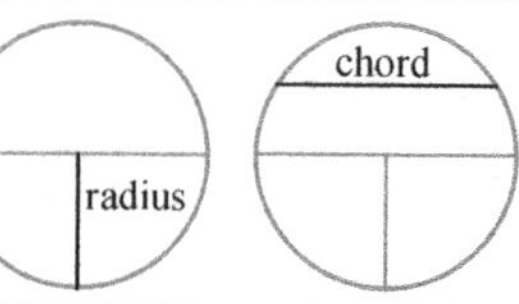

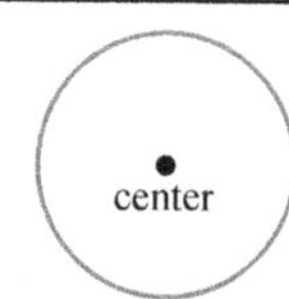

The length of the diameter is twice that of the radius.

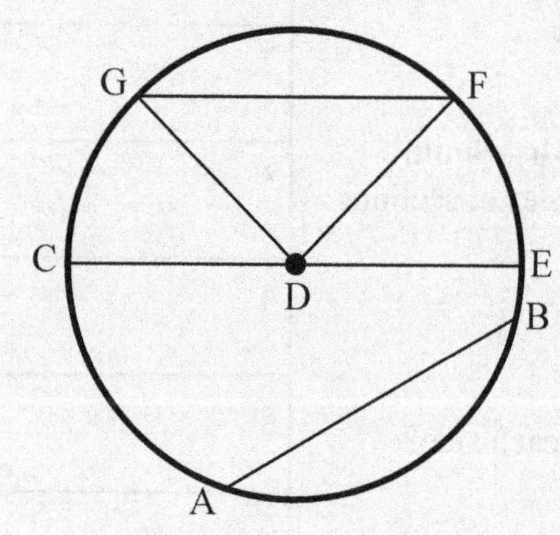

Use the figures to answer the following:

S1. What part of the circle is $\overline{\mathbf{CE}}$?

S2. Name two chords in circle **B**.

1. What part of circle A is $\overline{\mathbf{DF}}$?
2. What part of circle B is $\overline{\mathbf{VT}}$?
3. Name 3 radii in circle A.
4. Name 2 chords in circle A.
5. If the length of $\overline{\mathbf{CE}}$ is 16 ft., what is the length of $\overline{\mathbf{CD}}$?
6. Name the center of circle B.
7. Name two chords in circle B.
8. If $\overline{\mathbf{PS}}$ in circle B is 24 ft., what is the length of $\overline{\mathbf{XS}}$?
9. Name two radii in circle B.
10. Name the diameter in circle B.

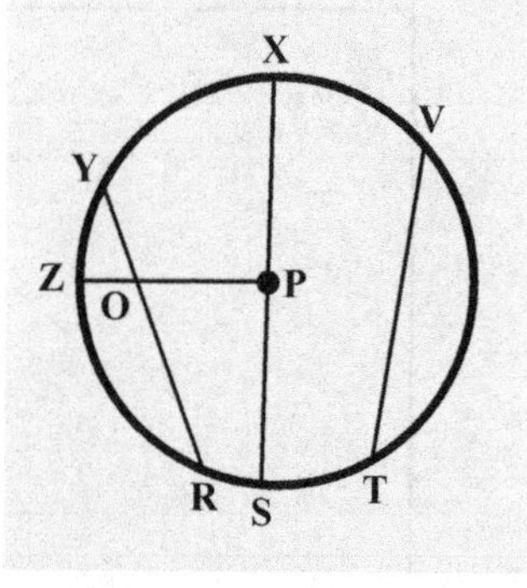

1.
2.
3.
4.
5.
6.
7.
8.
9.
10.
Score

Problem Solving

A city is in the shape of a square. If the perimeter is 52 miles, what is the length of each side of the city?

Review Exercises

1. If the radius of a circle is 16 ft., what is the length of the diameter?

2. The perimeter of a regular octagon is 72 ft. What is the length of each side?

3. $\begin{array}{r} 3.14 \\ \times \ 6 \\ \hline \end{array}$

4. $\frac{22}{7} \times 21 =$

5. Find the missing angle.

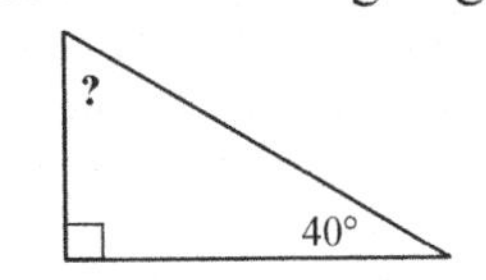

6. Classify sides ______
angles______

60°
5 cm.
3 cm.
4 cm.
30°

Helpful Hints

Use what you have learned to solve the following questions.

Circle A

H B C G A D E F

Use the figures to answer the following:

S1. What part of the circle is $\overline{\textbf{TW}}$?

S2. Name two radii in circle **A**.

1. What part of circle A is $\overline{\textbf{CD}}$?

2. What part of circle B is $\overline{\textbf{RS}}$?

3. Name 2 chords in circle A.

4. Name a diameter circle B.

5. If $\overline{\textbf{WT}}$ is 18 ft., what is the length of $\overline{\textbf{XT}}$?

6. If $\overline{\textbf{HE}}$ is 96 inches, what is the length of $\overline{\textbf{AB}}$?

7. Name two radii in circle B.

8. Name 2 chords in circle B.

9. Name the center of circle A.

10. Name another segment in circle A which is the same length as $\overline{\textbf{AB}}$.

Circle B

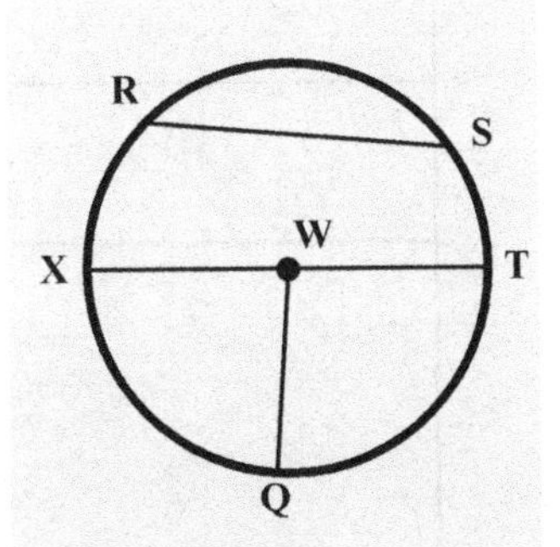

1.
2.
3.
4.
5.
6.
7.
8.
9.
10.
Score

Problem Solving

A fence was built around a yard in the shape of a square with sides of 14 ft. If materials cost $12 per foot, how much did the fence cost?

Review Exercises

1. If the radius of a circle is 12 ft., what is the length of the diameter?

2. $\frac{22}{7} \times 35 =$

3. 3.14×8

4. The diameter of a circle is 56 inches. What is the length of the radius?

5. What kind of triangle has all sides of a different length?

6. What kind of triangle has two sides the same length?

Helpful Hints

The distance around a circle is called its **circumference**. The Greek letter π = pi = 3.14 or $\frac{22}{7}$. To find the **circumference**, multiply π × **diameter**. **Circumference = π × d**

Examples:

$C = \pi \times d$
$C = 3.14 \times 6$

(circle with diameter 6 ft.)

$3.14 \times 6 = 18.84$ ft.

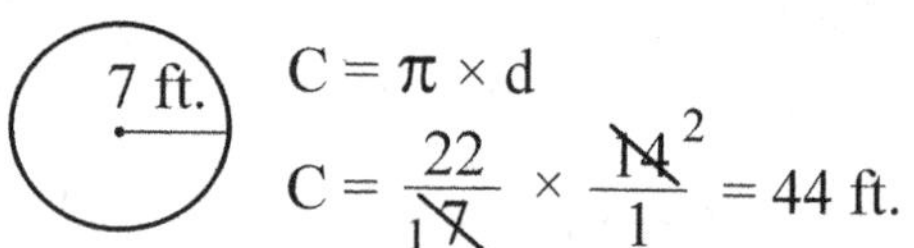

$C = \pi \times d$

$C = \frac{22}{\cancel{7}_1} \times \frac{\cancel{14}^2}{1} = 44$ ft.

(Hint: If the diameter is divisible by 7, use $\pi = \frac{22}{7}$)

Find the circumference of each of the following.

S1. 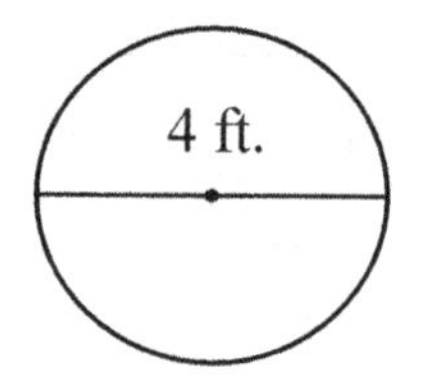

S2. 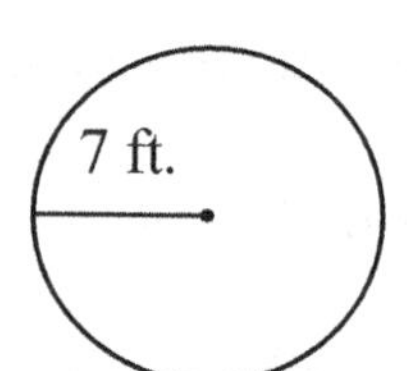

1. 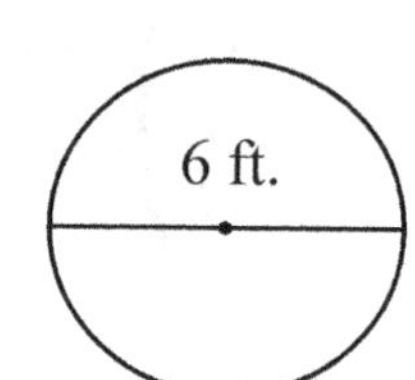

2. 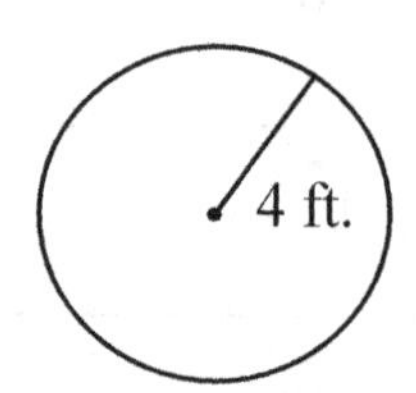

3. A circle with diameter 9 ft.

4. A circle with radius 14 ft.

5. 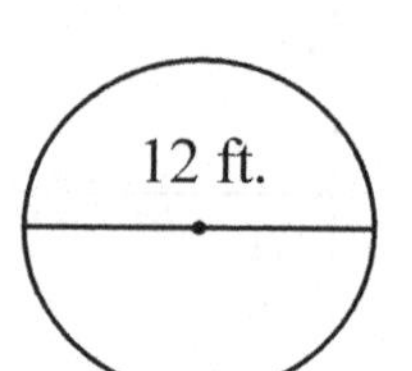

6. 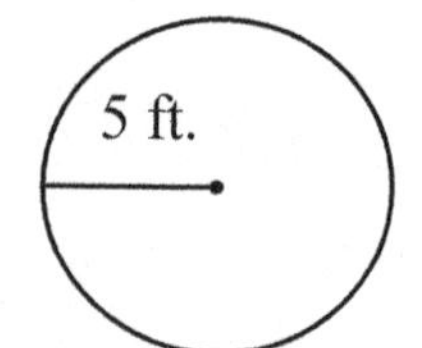

7. A circle with radius 2 ft.

1.
2.
3.
4.
5.
6.
7.

Score

Problem Solving

Mr. Vargas earned $67,200 last year.
What were his average monthly earnings?

Review Exercises

For 1-6, find the perimeter for each figure. Draw a sketch.

1. A square with sides of 73 inches.
2. A rectangle with length 32 ft., and width 18 ft.
3. A regular pentagon with sides 63 meters.
4. An equilateral triangle with sides 29 ft.
5. A regular octagon with sides 46 ft.
6. A regular hexagon with sides 112 ft.

Helpful Hints

Use what you have learned to solve the following problems.

1. If there is no figure, draw a sketch.
2. If the diameter is divisible by 7, use $\pi = \frac{22}{7}$

Find the circumference of each of the following.

S1.

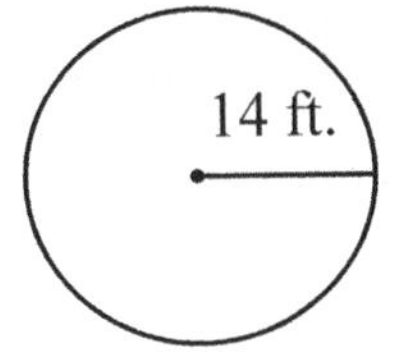

S2.

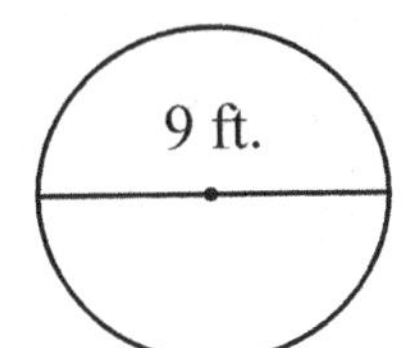

1.

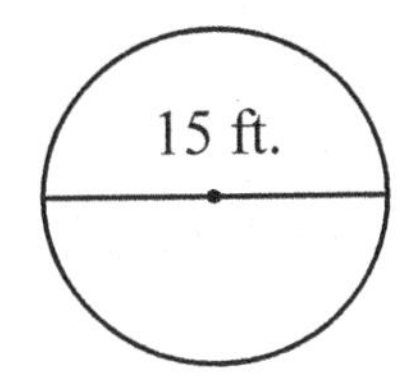

2.

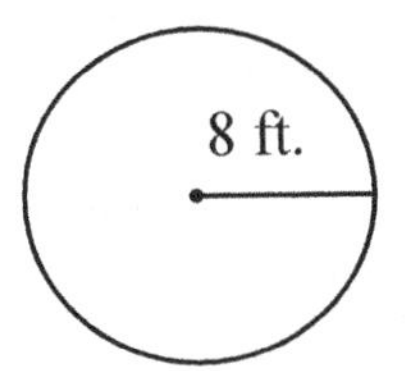

3.

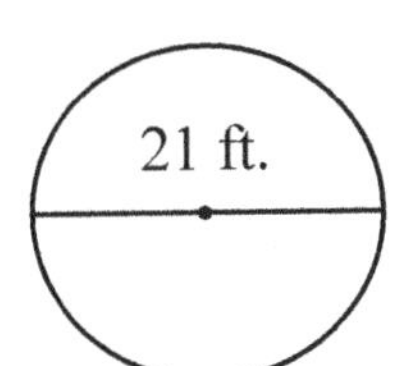

4. A circle with radius 21 ft.

5. A circle with diameter 35 ft.

6.

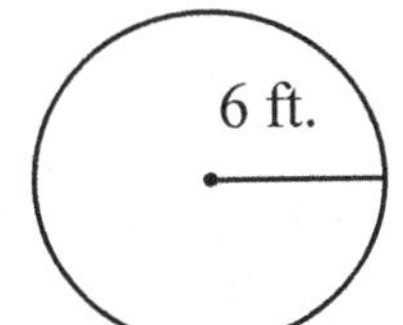

7.

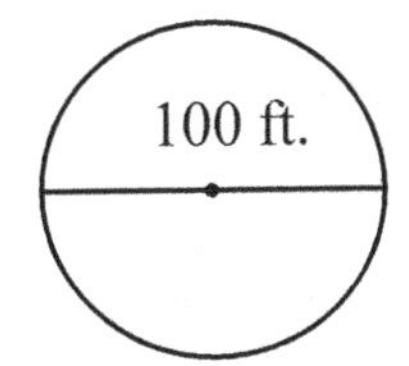

1.
2.
3.
4.
5.
6.
7.
Score

Problem Solving

A fountain is in the shape of a circle. If its radius is 14 ft., how far is it all the way around the fountain?

Review Exercises

1. Find the circumference.

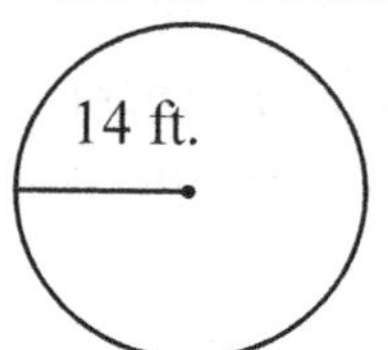

2. Find the perimeter.

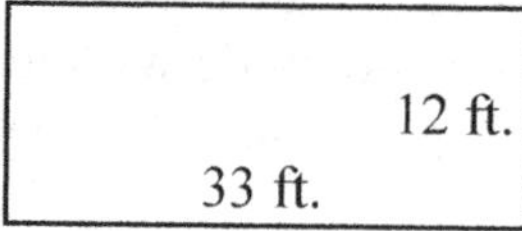

3. Find the circumference

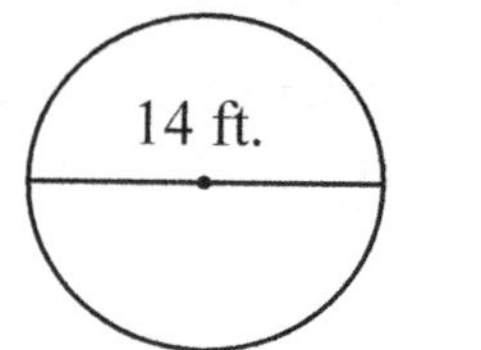

4. $3.14\overline{)25.12}$

5. $66 \div \frac{22}{7} =$

6. Classify by angles.

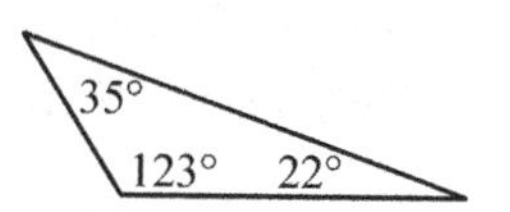

Helpful Hints

Use what you have learned to solve the following problems.

1. Read the problem carefully to understand what is being asked, draw a sketch.
2. If the diameter is divisible by 7, use $\pi = \frac{22}{7}$ * diameter = C ÷ π

S1. A circular dodge ball court has a diameter of 35 ft. How far is it all the way around the court?

S2. Mrs. Isaac is making cloth emblems for her club's jackets. If each emblem has a radius of 7 cm., what is the circumference of each emblem?

1. Tara has a frisbee with a diameter of 14 inches. What is the circumference of the frisbee?

2. A city is surrounded by a road in the shape of a circle. If the radius of the road is 14 miles, how far would you drive to make a complete circle around the city?

3. Circle A has diameter 21 ft. and Circle B has diameter 28 ft. How much greater is the circumference of Circle B than Circle A?

4. The circumference of a circle is 12.56 ft. What is the diameter? (Use π = 3.14)

5. The circumference of a circle is 18.84 inches. What is the radius? (Use π = 3.14)

6. Find the circumference.

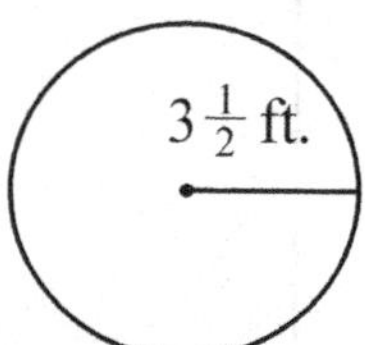

7. Find the circumference.

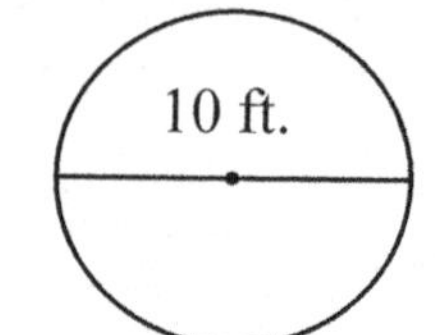

1.
2.
3.
4.
5.
6.
7.
Score

Problem Solving

The perimeter of a square is 496 ft. What is the length of each side?

Review Exercises

1. What is the complement of 23°?

2. What is the supplement of 97°?

3. What polygon has one pair of parallel sides?

4. $88 \div \frac{22}{7} =$

5. $3.14\overline{)37.68}$

6. Two angles of a triangle are 37° and 72°. What is the measure of the third angle?

Helpful Hints

Use what you have learned to solve the following problems.

* Draw a sketch.
* Decide which is easier, using $\pi = 3.14$ or $\pi = \frac{22}{7}$
* $d = C \div \pi$

S1. The circumference of a circle is 110 ft. What is the diameter? (Use $\pi = \frac{22}{7}$)

S2. The circumference of a circle is 47.1 ft. What is the diameter?

1. The radius of a circular shaped fountain is 14 ft. What is the diameter of the fountain?

2. A race track in the shape of a circle has a radius of 56 meters. What is the circumference of the race track?

3. From the north pole through the center of the earth to the south pole is approximately 8,000 miles. What is the distance around the equator?

4. If the circumference of a circle is 264 ft., what is the diameter? ($\pi = \frac{22}{7}$)

5. If the circumference of a circle is 62.8 ft., what is the diameter? ($\pi = 3.14$)

6. Find the circumference.

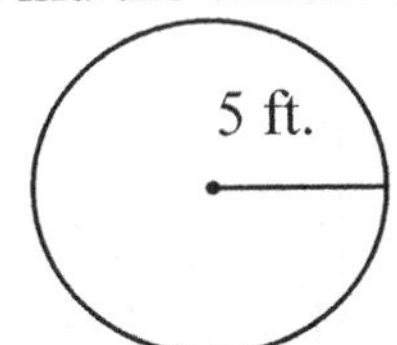

7. Find the circumference.

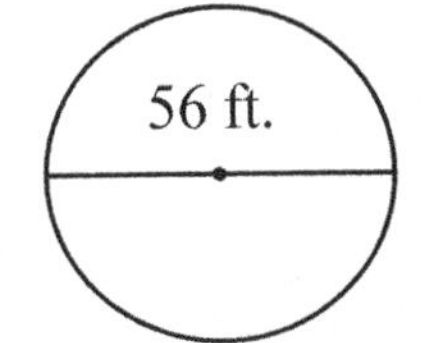

1.
2.
3.
4.
5.
6.
7.
Score

Problem Solving

A sports court is in the shape of a rectangle with length 84 ft. and width 36 ft. What is the perimeter of the sports court?

Review Exercises

1. Find the perimeter of a regular decagon with sides of 17 ft.

2. Find the circumference

6 ft.

3. Find the perimeter.

28 ft.
68 ft.

4. Find the circumference.

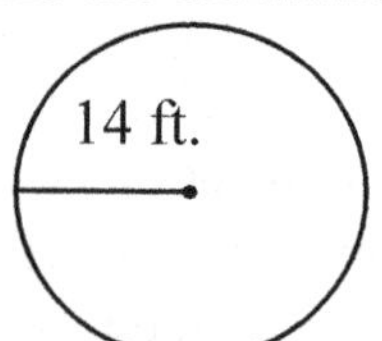

5. Find the perimeter of an equilateral triangle with sides of 17 ft.

6. Find the perimeter of a square with sides 19 ft.

Helpful Hints

Use what you have learned to solve the following problems.

* Refer to helpful hints on previous pages if necessary.

Find the perimeter or circumference of the following polygons.

S1. 18 ft. 14 ft. 10 ft. 12 ft.

S2. 24 ft. 16 ft. 16 ft. 32 ft.

1. 21 ft.

2. 2 ft. 12 ft.

3. 29 ft.

4. 6 ft. 3 ft. 4 ft. 5 ft. 3 ft.

5. 9 ft. 6 ft. 6 ft. 15 ft. 15 ft.

6. 4 ft.

7. 59 ft.

8. 16 ft. 8 ft.

9. 152 ft. 78 ft.

10. 5 ft.

1.
2.
3.
4.
5.
6.
7.
8.
9.
10.
Score

Problem Solving

A family is on a vacation and has to drive 385 miles. If the average speed of their car is 55 miles per hour, how long will it take to reach their destination?

Review Exercises

1. Classify 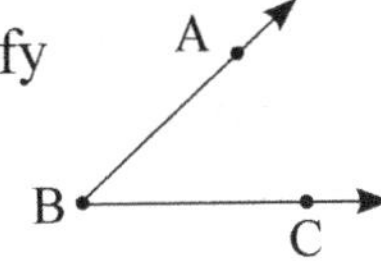

2. Classify 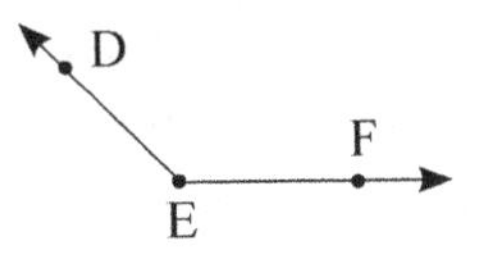

3. Classify 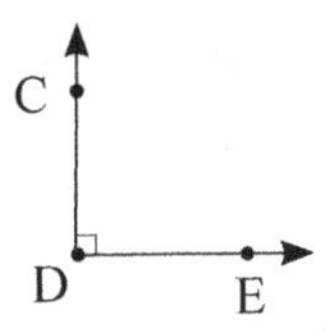

4. How many sides does an octagon have?

5. What is a five-sided polygon called?

6. What is a ten-sided polygon called?

Helpful Hints

Use what you have learned to solve the following problems.
*Draw a sketch and label the parts.
*Refer to "Helpful Hints" on previous pages if necessary.

Find the perimeter or circumference of the following. Draw a sketch.

S1. **Square:** Sides, 29 ft.

S2. **Circle:** Radius, 6 ft.

1. **Rectangle:** Length, 36 ft. Width, 19 ft.

2. **Equilateral Triangle:** Sides, 86 ft.

3. **Circle:** Diameter, 28 ft.

4. **Regular Pentagon:** Sides, 37 ft.

5. **Square:** Sides, 115 ft.

6. **Regular Decagon:** Sides, 39 ft.

7. **Rectangle:** Length, 46 ft. Width, 96 ft.

8. **Circle:** Radius, 21 ft.

9. **Regular Hexagon:** Sides, 42 ft.

10. If the perimeter of a square is 528 ft., what are the lengths of the sides?

1.
2.
3.
4.
5.
6.
7.
8.
9.
10.
Score

Problem Solving

During a week Julio started with 550 dollars. First he spent 136 dollars, then he earned 208 dollars, and finally he spent 316 dollars. How much money did he have left?

Review Exercises

1. Draw two lines $\overleftrightarrow{CD}$ and $\overleftrightarrow{FG}$, that are perpendicular.

2. Draw two lines $\overleftrightarrow{LM}$ and $\overleftrightarrow{RS}$, that are parallel.

3. What angle is complementary to 15°?

4. What angle is supplementary to 15°?

5. In a right isosceles triangle, one angle measures 90°. What are the measures of the other two angles?

6. What is the measure of ∠ **ABC**?

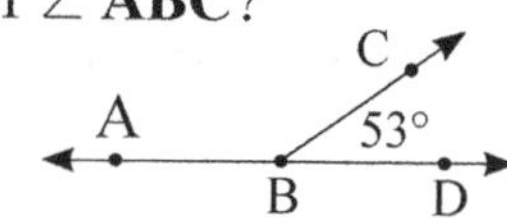

Helpful Hints

Use what you have learned to solve the following problems.

*Draw a sketch.

*Refer to previous pages if necessary.

S1. What is the circumference of a circular sports court if the diameter is 35 ft?

S2. A flower garden is in the shape of a regular pentagon. If its sides are 17 ft., what is the perimeter of the flower garden?

1. A coach wants to paint a white line around the perimeter of a rectangular court. If the width is 35 ft. and the length is 70 ft., what is the perimeter of the court?

2. The circumference of a circle is 154 ft. What is the diameter? ($\pi = \frac{22}{7}$)

3. The sides of a square are 180 ft. What is the perimeter?

4. The perimeter of a regular hexagon is 228 ft. What are the lengths of the sides?

5. A man wants to build a fence around a yard in the shape of a square with sides of 24 ft. If fencing costs 15 dollars per foot, what will be the cost of the fence?

6. The perimeter of a square is 624 ft. What are the lengths of the sides?

7. What is the perimeter of a regular octagon with sides of 23 ft.?

1.
2.
3.
4.
5.
6.
7.
Score

Problem Solving

A tank holds 55,000 gallons. If 12,500 gallons were removed one day, and 13,450 gallons the next day, how many gallons are left?

Review Exercises

For 1-6, find the perimeter or circumference. (Draw a sketch.)

1. **Square:** Sides, 16 ft.

2. **Rectangle:** Length, 26 ft. Width, 19 ft.

3. **Circle:** Radius, 3 ft.

4. **Circle:** Diameter, 21 ft.

5. **Regular Pentagon:** Sides, 28 ft.

6. **Equilateral Triangle:** Sides, 33 ft.

Helpful Hints

Use what you have learned to solve the following problems.

S1. What is the diameter of a circle with circumference 198 ft.? ($\pi = \frac{22}{7}$)

S2. What is the radius of a circle with circumference 18.84 ft.? ($\pi = 3.14$)

1. A circular race track has a circumference of 3 miles. In a 375 mile car race, how many laps around the track will each auto travel?

2. The walls of the Pentagon Building are 921 ft. long. What is the perimeter of the building? (Hint: The building is in the shape of a regular pentagon.)

3. Each side of a square is 4 ft. long. How many inches is it all the way around the square?

4. The perimeter of a rectangle is 60 ft. If the length is 18 ft., what is the width?

5. The diameter of a track is 7 miles. How many laps must a car drive to cover a distance of 88 miles?

6. What is the perimeter of a regular octagon with sides 38 ft.?

7. Circle A has a diameter of 14 ft. Circle B has a diameter of 35 ft. How much longer is the circumference of Circle B than Circle A?

1.
2.
3.
4.
5.
6.
7.
Score

Problem Solving

A family planned a 750 mile trip. They drove 335 miles the first day and 255 miles the second day. How many more miles must they drive?

Review Exercises

1. List three types of polygons.

2. List the four types of angles.

3. List four parts of a circle.

4. List the three types of triangles classified by sides.

5. List three types of triangles classified by angles.

6. List four polygons that have at least five sides.

Helpful Hints

The Pythagorean Theorem: In a right triangle the sum of the squares of the sides is equal to the square of the hypotenuse.
* You will need a calculator with a "$\sqrt{\ }$" key.

Hypotenuse → c, a, b

The Pythagorean Theorem:
$$a^2 + b^2 = c^2$$

Example: Find the hypotenuse.

c, a(6 ft.), b(5 ft.)

$a^2 + b^2 = c^2$

$(6 \times 6) + (5 \times 5) = c^2$

$36 + 25 = c^2$

$61 = c^2$

$\sqrt{61} = c$

$\sqrt{61} = 7.810$

Round to the nearest whole number. $c = 8$

Find the hypotenuse of each triangle using the formula $a^2 + b^2 = c^2$. Round answers to the nearest whole numbers.

S1.

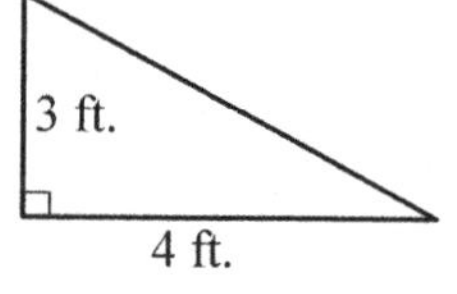

S2.

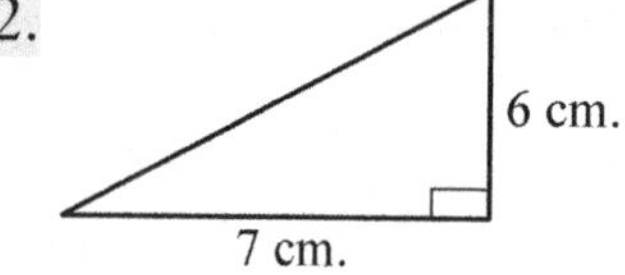

1.

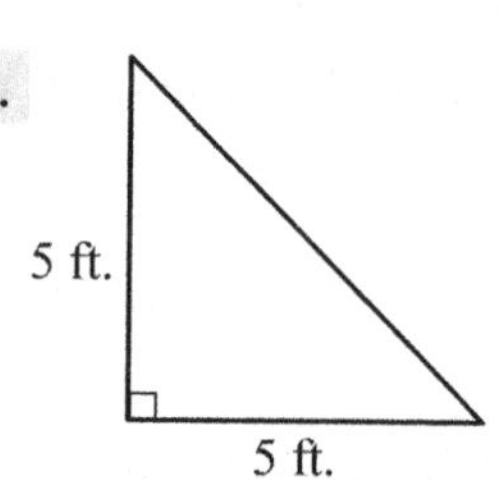

2.

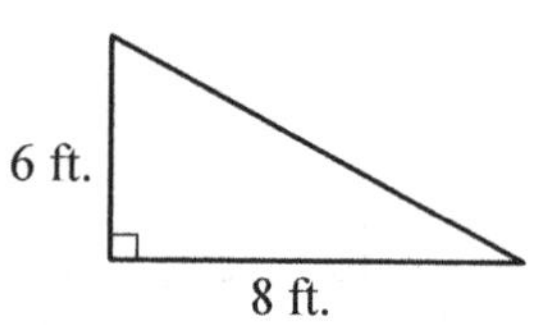

3.

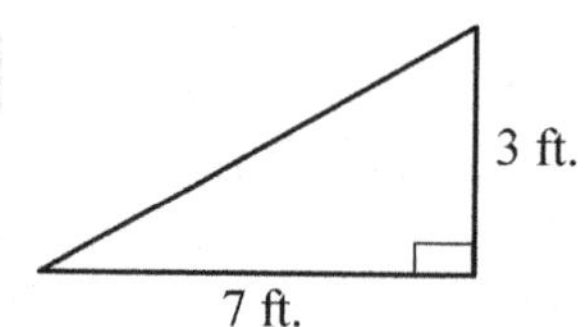

4.

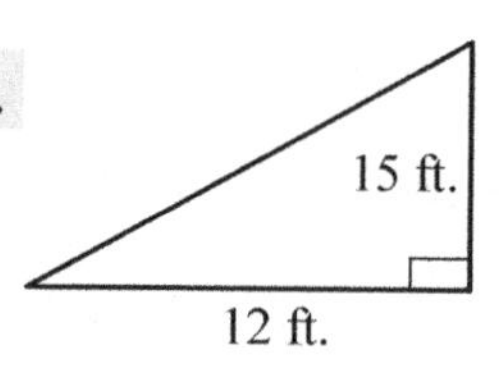

5.

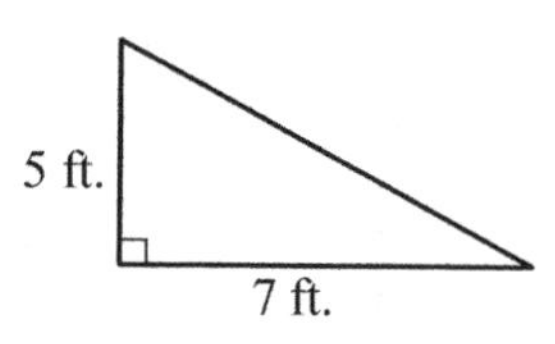

6.

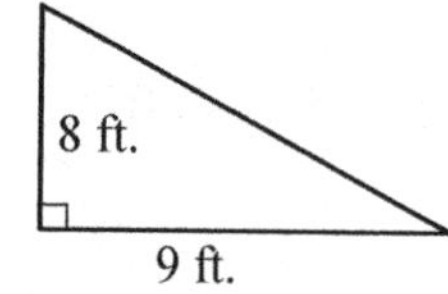

7.

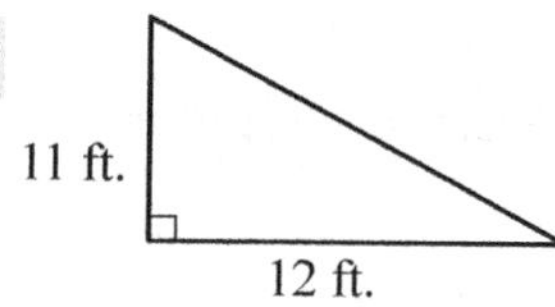

1.

2.

3.

4.

5.

6.

7.

Score

Problem Solving

Three pounds of steak costs $9.75. One pound of fish is $2.55. How much will two pounds of steak and three pounds of fish cost?

Review Exercises

For 1-6, find the perimeter or circumference. (Draw a sketch.)

1. **Square:** Sides, 17 ft.
2. **Rectangle:** Length, 19 ft. Width, 7 ft.
3. **Regular Hexagon:** Sides, 29 ft.
4. **Circle:** Radius, 6 ft.
5. **Circle:** Diameter, 7 ft.
6. **Equilateral Triangle:** Sides, 175 ft.

Helpful Hints

The **hypotenuse** of a right triangle may be found using the Pythagorean Theorem. $a^2 + b^2 = c^2$

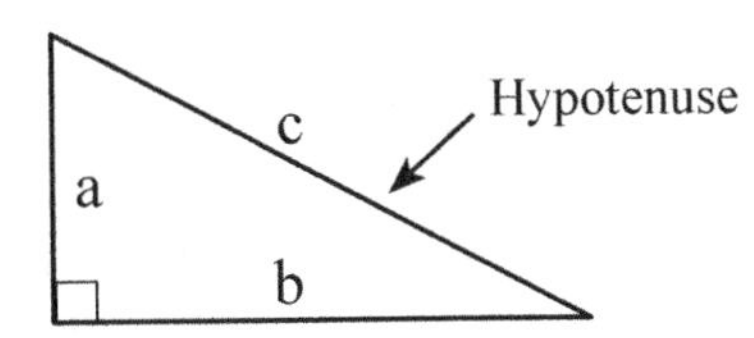

* You will need a calculator with a " $\sqrt{\ }$ " key.

Example: Find the hypotenuse in the triangle below.

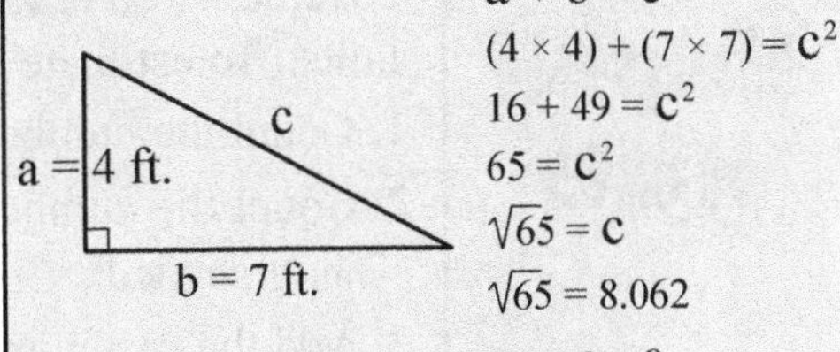

$a^2 + b^2 = c^2$
$(4 \times 4) + (7 \times 7) = c^2$
$16 + 49 = c^2$
$65 = c^2$
$\sqrt{65} = c$
$\sqrt{65} = 8.062$

Round to the nearest whole number. $c = 8$

Find the hypotenuse of each triangle using the formula $a^2 + b^2 = c^2$.
Round answers to the nearest whole numbers.

S1.

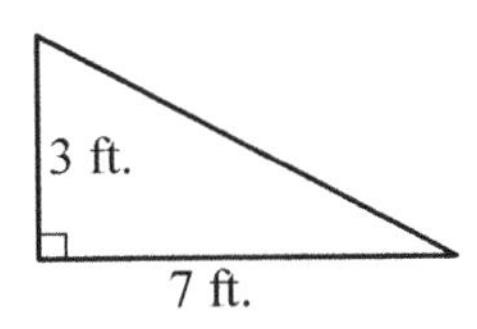

S2.

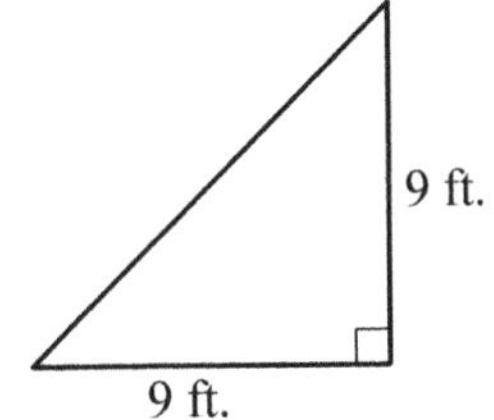

1.

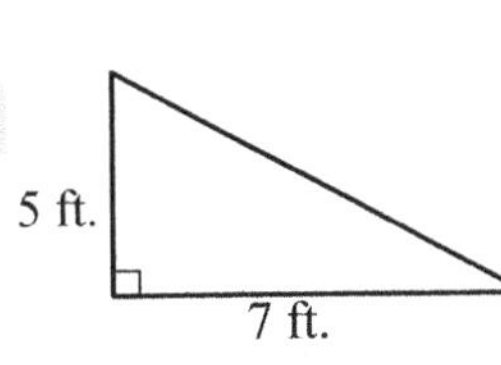

2.

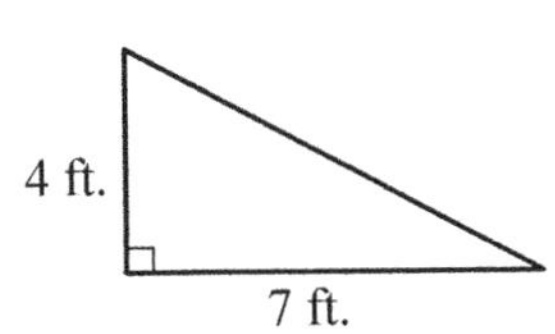

3.

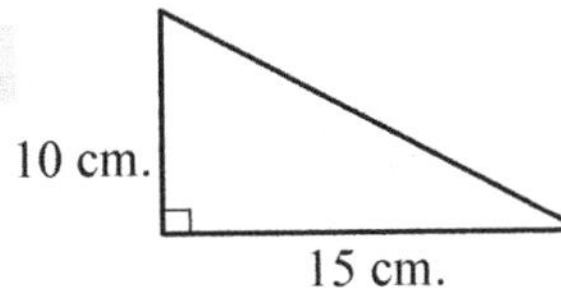

4.

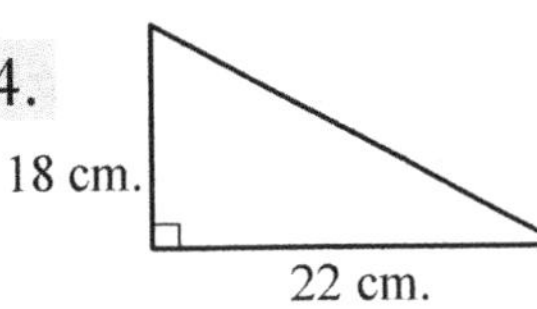

5.

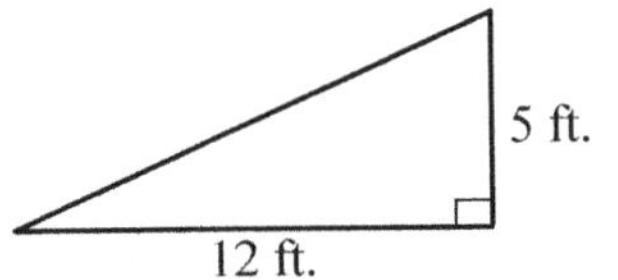

6.

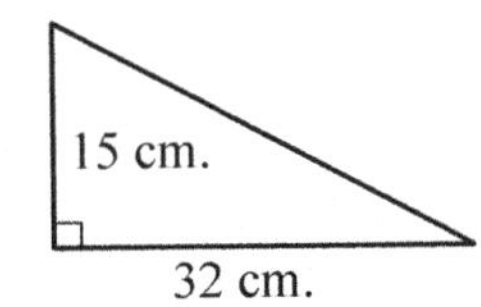

7.

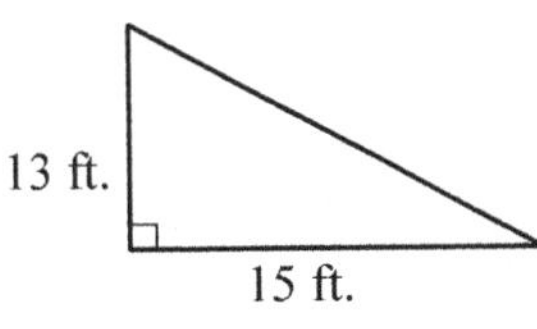

1.
2.
3.
4.
5.
6.
7.
Score

Problem Solving

A farmer wants to take 2,760 eggs to the market. If he packs the eggs in cartons which hold 12 eggs, how many cartons does he need.

Review Exercises

1. Find the perimeter.

9 ft. | 25 ft.

2. Find the perimeter.

119 ft.

3. Find the perimeter.

42 ft.

4. Find the circumference.

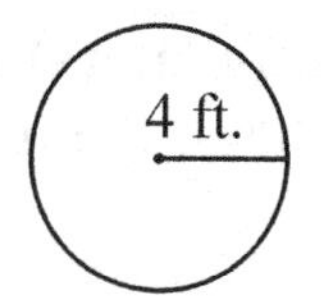

5. How many sides does a hexagon have?

6. How many sides does an octagon have?

Helpful Hints

Area can be thought of as the amount of surface covered by an object. It is expressed in square units. To estimate area:

1. Count the number of squares completely shaded.

2. Count the number of squares more than half shaded.

3. Add the two numbers.

Example:

$$\begin{array}{r} 6 \\ +\ 3 \\ \hline 9 \end{array}$$

The area is about 9 ft.^{2}

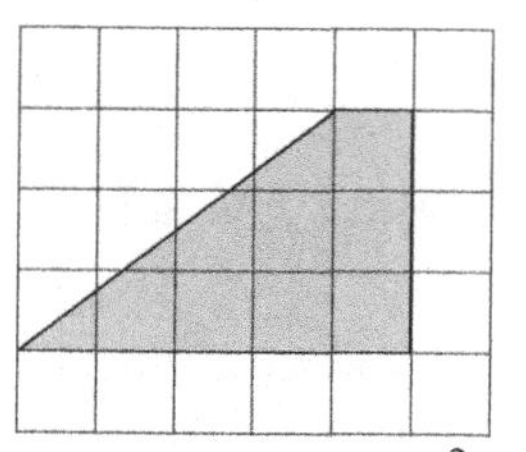

Each square is 1 ft.^{2}
ft.^{2} = square feet

Estimate the area of each shaded figure. Each square is 1 ft.^{2}.

S1.

S2.

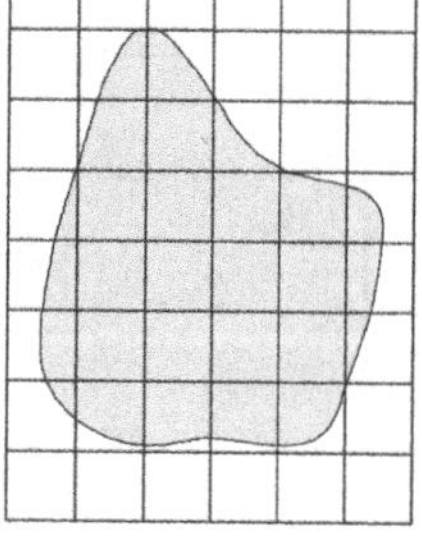
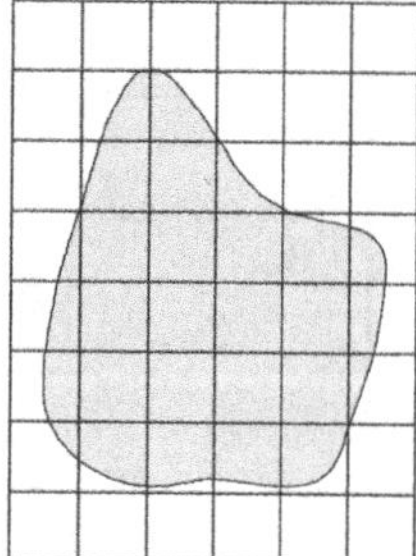

1.

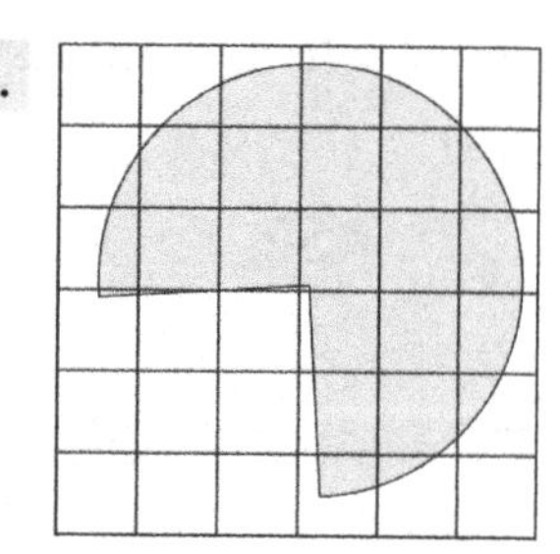

2.

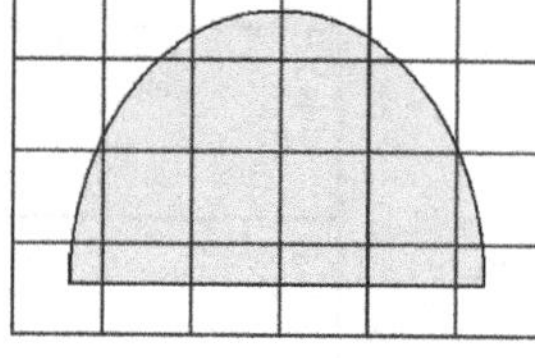
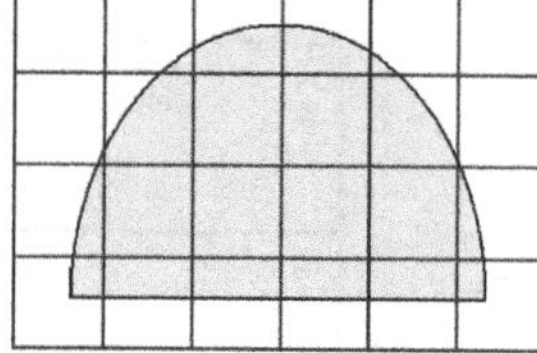

3.

4.

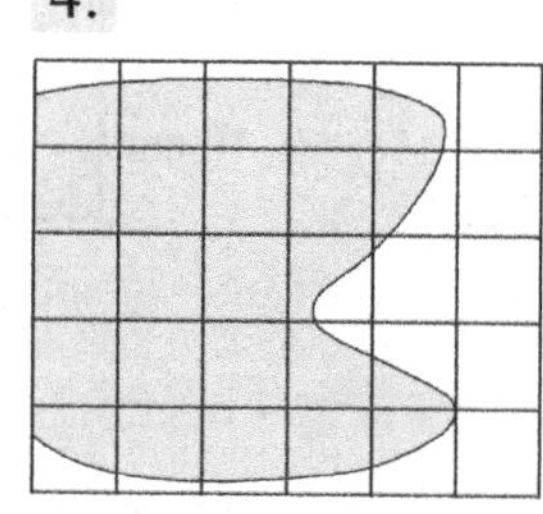

5.

6.

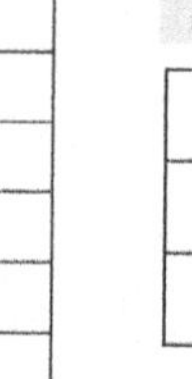

7.

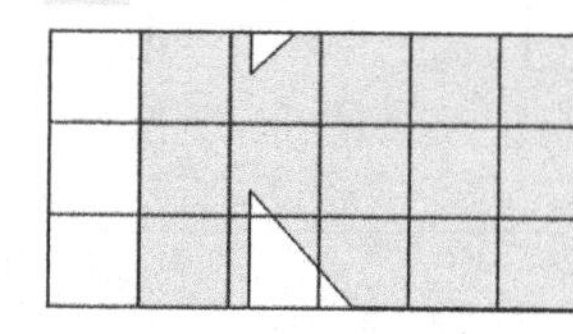

1. ______
2. ______
3. ______
4. ______
5. ______
6. ______
7. ______
8. ______
9. ______
10. ______

Problem Solving

Pat's test scores were 90, 87, 93, 95, and 95. What was her average test score?

Score

Review Exercises

1. Classify by sides.

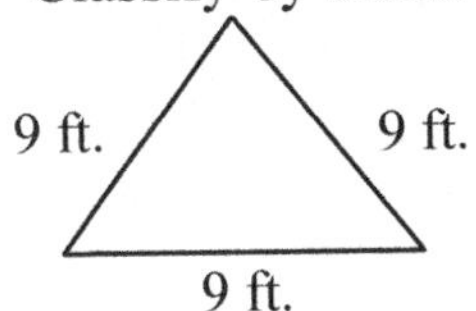

2. Classify by angles.

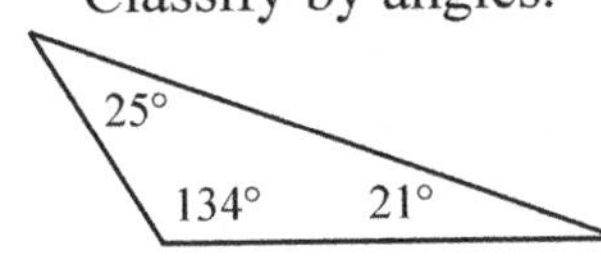

3. Find the hypotenuse. (Round to the nearest whole number.)

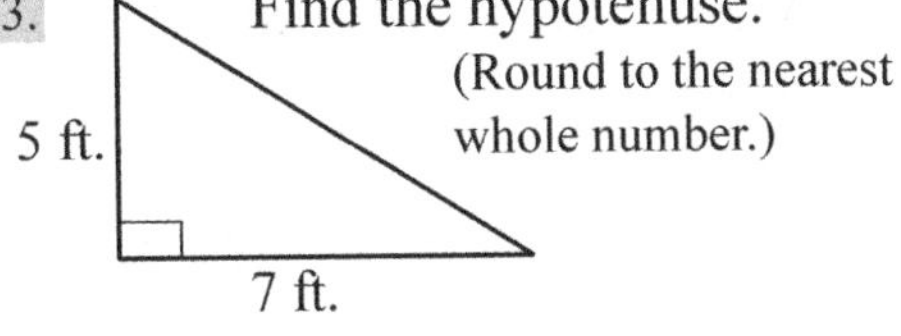

4. Find the circumference.

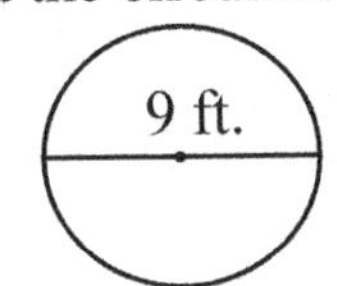

5. Find the circumference.

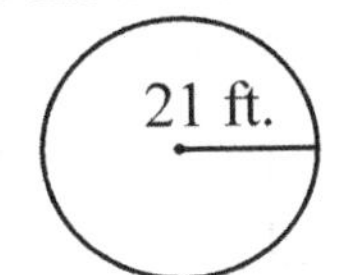

6. Find the perimeter.

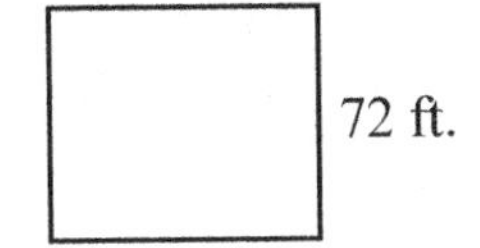

Helpful Hints

Use what you have learned to solve the following problems.

Shade in the figure the area given. Each square is 1 ft^2.

S1. 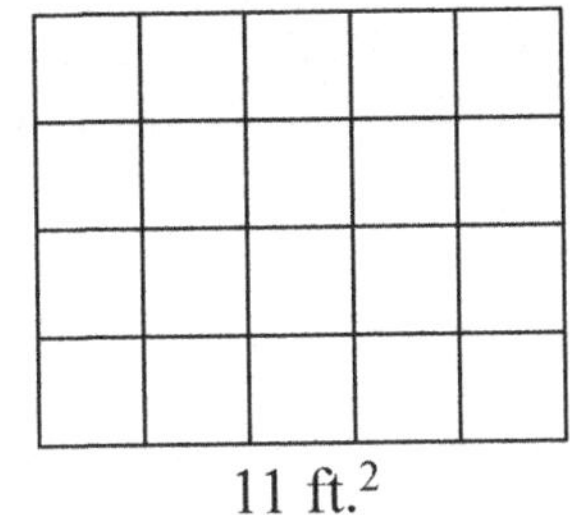

11 ft.^{2}

S2. 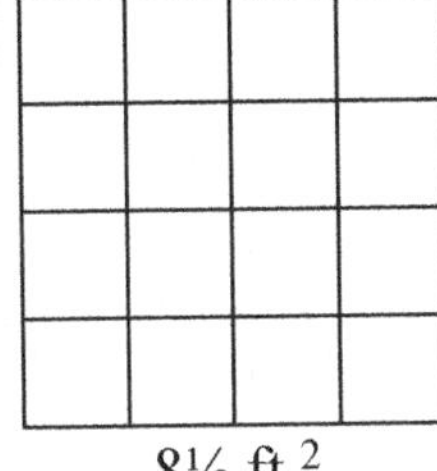

8½ ft.^{2}

1. 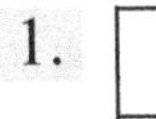 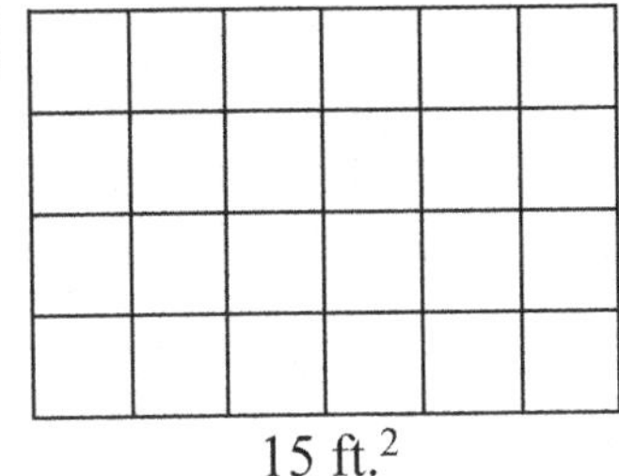

15 ft.^{2}

2. 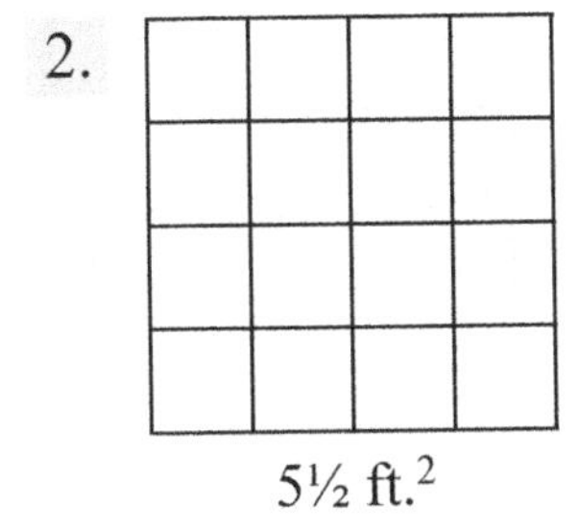

5½ ft.^{2}

3. 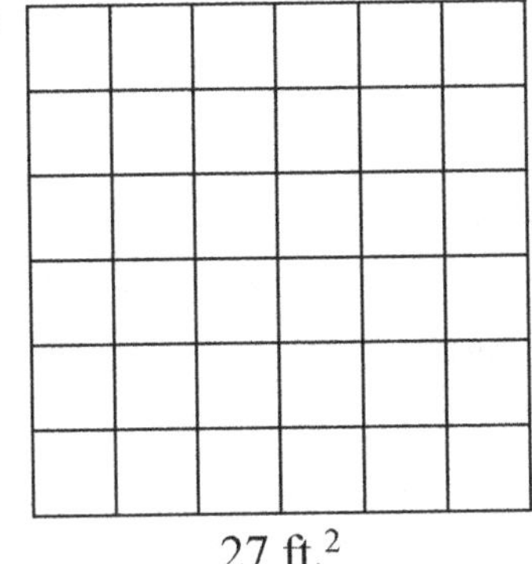

27 ft.^{2}

4.

9½ ft.^{2}

5.

7 ft.^{2}

6. 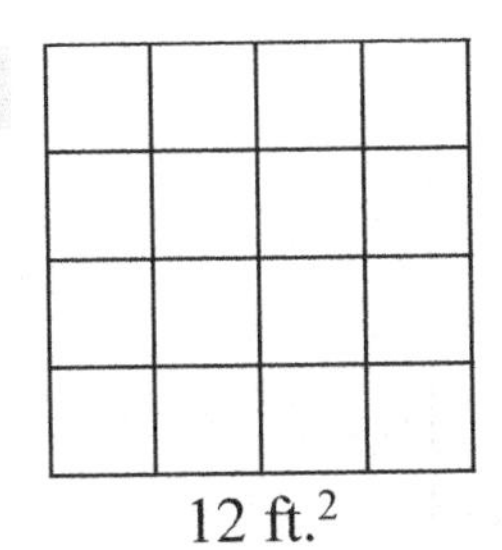

12 ft.^{2}

7. 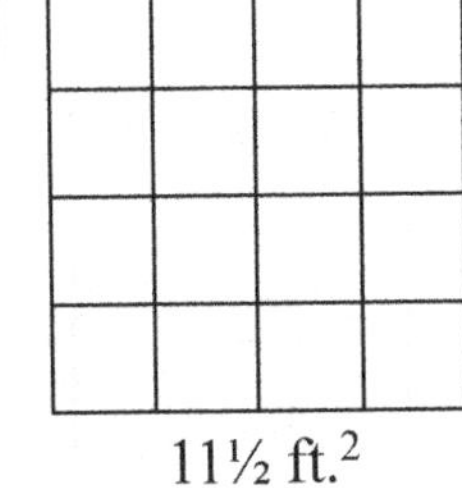

11½ ft.^{2}

1.
2.
3.
4.
5.
6.
7.
8.
9.
10.

Score

Problem Solving

Sam bought a new car. He made a down payment of 2,000 dollars, and will make 48 monthly payments of 400 dollars. What is the total cost of the car?

Review Exercises

1. What is the complement of 6°?

2. What is the supplement of 17°?

3. Name three polygons with two pairs of parallel sides.

4. If the perimeter of a square is 580 ft., find the length of its sides.

5. Find the perimeter.

16 ft.

6. Find the circumference.

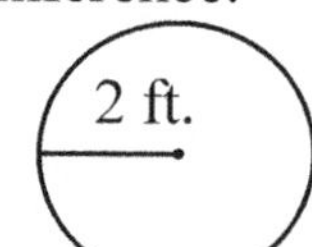

Helpful Hints

The number of square units needed to cover a region is called its area. Formulas are used to find areas.

Examples:

area square = side × side

s = 7 ft.

$A = s \times s$

$A = 7 \times 7$

$$\begin{array}{r} 7 \\ \times\ 7 \\ \hline 49 \text{ sq. ft.} \end{array}$$

area rectangle = length × width

w = 7 ft.

l = 12 ft.

$A = l \times w$

$A = 12 \times 7$

$$\begin{array}{r} 12 \\ \times\ 7 \\ \hline 84 \text{ sq. ft.} \end{array}$$

Find the following areas. Write the formula. Next, substitute values. Then solve the problem. If there is no diagram, draw a sketch.

S1. 13 ft.

S2.

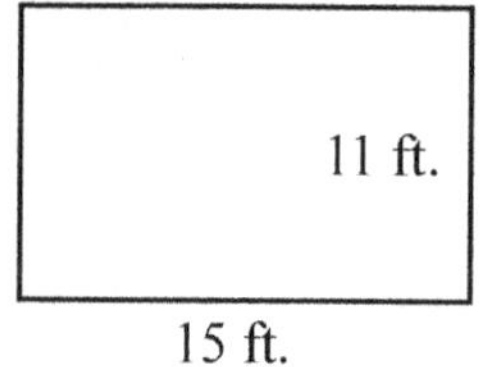

1. 6 ft. 14 ft.

2.

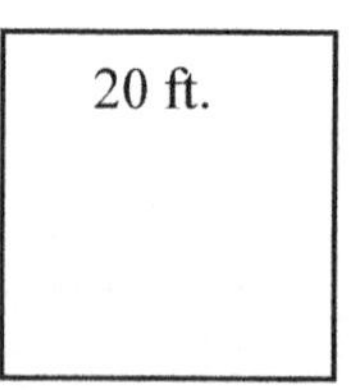

3. A rectangle with length 12 ft. and width 11 ft.

4. 17 ft. 23 ft.

5. 4.5 ft 3 ft.

6. 25 ft.

7. A square with sides 180 ft.

1.

2.

3.

4.

5.

6.

7.

Score

Problem Solving

A school has 32 classrooms with 25 students in each class. If the school has 42 teachers, what is the total of all students and teachers?

Review Exercises

1. Find the area.

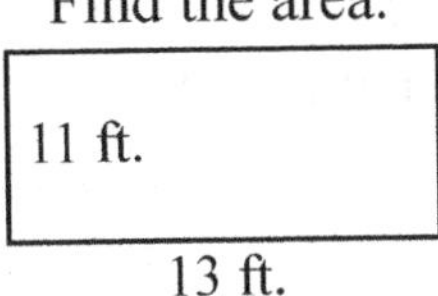

2. Find the area.

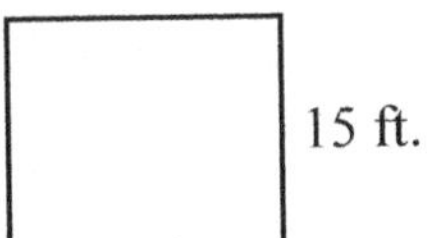

3. Find the perimeter.

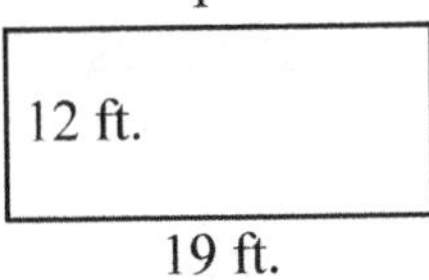

4. Find the perimeter.

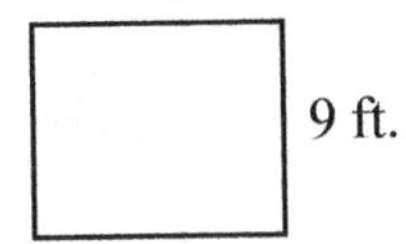

5. Find the circumference.

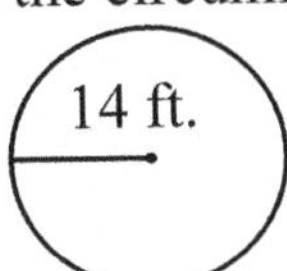

6. Find the circumference.

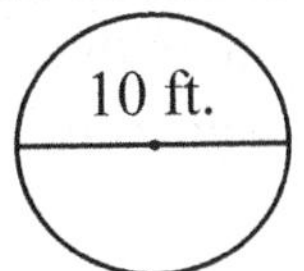

Helpful Hints

Use what you have learned to solve the following problems.

1. Start with the formula.
2. Substitute the values.
3. Solve the problem.

* Answers are in square units.
* sq. ft. = $ft.^2$

Square: $A = s \times s$

Rectangle: $A = l \times w$

Find the following areas. If there is no figure, make a sketch.

S1.

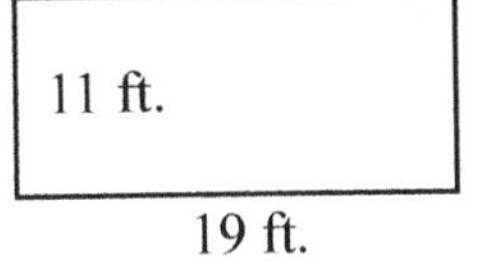

S2.

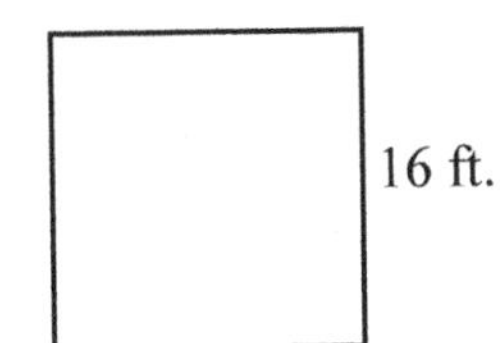

1.

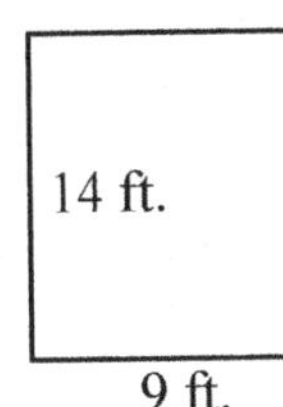

2. A square with sides 25 ft.

3. A rectangle with length 25 ft. and width 19 ft.

4.

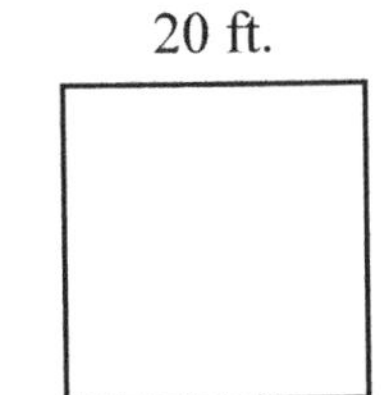

5. If the area of a square is 196 $ft.^2$, what are the lengths of its sides?

6. If a rectangle has an area of 400 $ft.^2$ and the width is 16 ft., what is the length?

7. 150 cm.
200 cm.

1.
2.
3.
4.
5.
6.
7.
Score

Problem Solving

Square A has sides of 6 ft. and Square B has sides of 8 ft. How much larger is the area of Square B than Square A?

Review Exercises

1. Find the perimeter of a square with sides of 17 ft.

2. Find the area of a square with sides of 17 ft.

3. Find the area of a rectangle with length 16 ft. and width 10 ft.

4. Find the perimeter of a rectangle with length 16 ft. and width 10 ft.

5. Find the hypotenuse. (Round to the nearest whole number.)

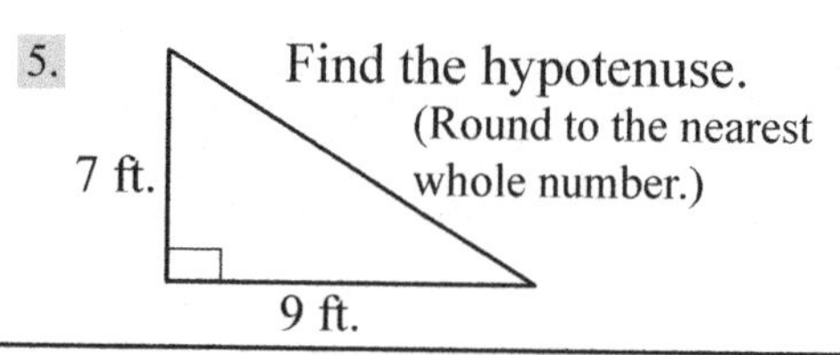

6. Two of the angles of a triangle are 72° and 58°, what is the third angle?

Helpful Hints

Area of a triangle $= \frac{\textbf{base} \times \textbf{height}}{\textbf{2}} = \frac{\textbf{b} \times \textbf{h}}{\textbf{2}}$

Examples:

height = 8 ft.

base = 7 ft.

$A = \frac{b \times h}{2}$

$A = \frac{7 \times 8}{2} = \frac{56}{2} = 2\overline{)56}$ (28 sq. ft.)

Area of a parallelogram = base × height = b × h

Example:

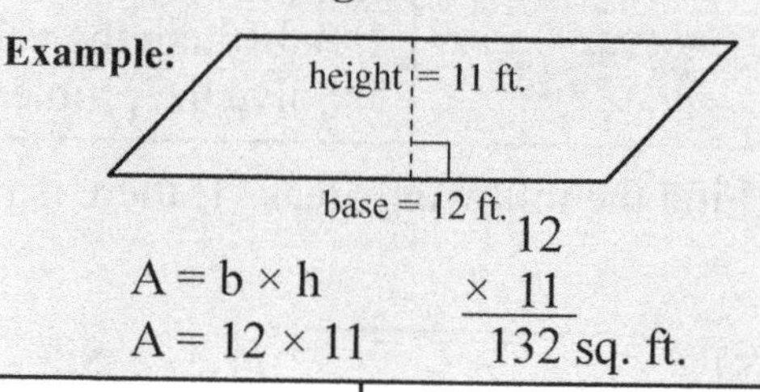

$A = b \times h$

$A = 12 \times 11$

$12 \times 11 = 132$ sq. ft.

Find the area of each of the following. Start with the formula. Substitute the values. Finally, solve the problem. If there is no diagram, make a sketch.

S1.

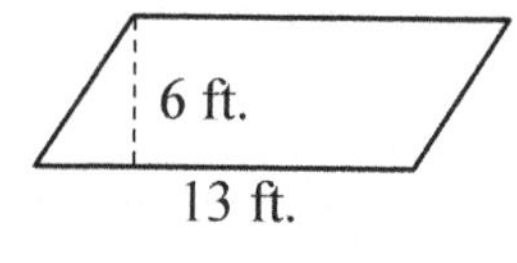

S2.

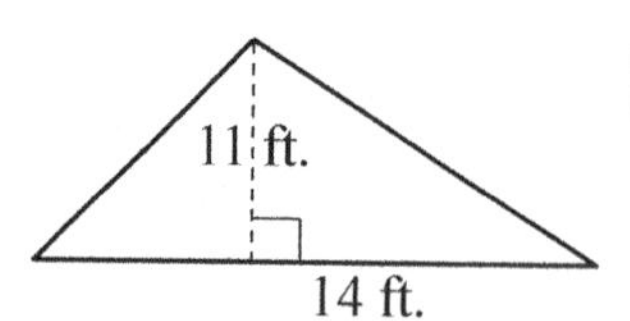

1.

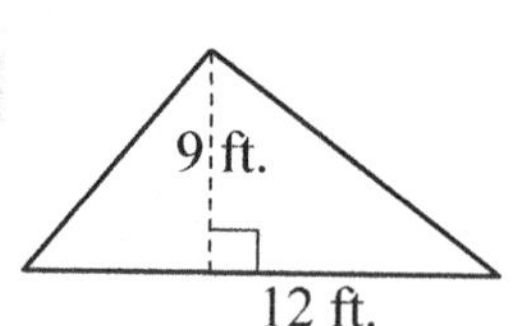

2.

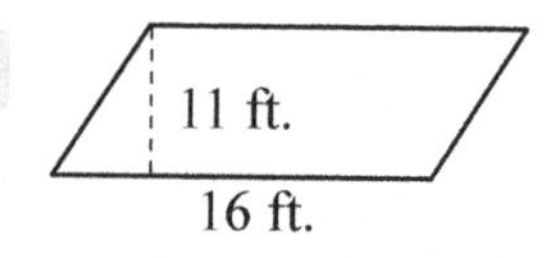

3. A triangle with base 5 ft. and height 7 ft.

4. A parallelogram with base 13 ft. and height 7 ft.

5.

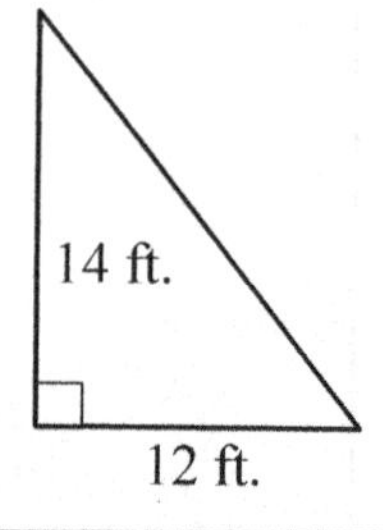

6.

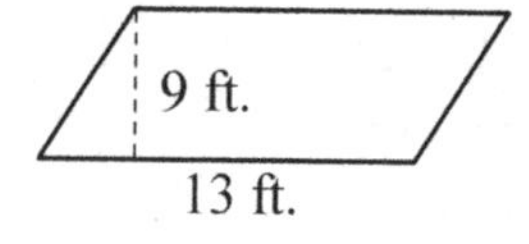

7.

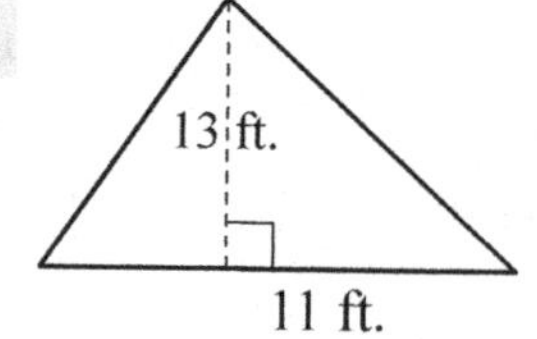

1.
2.
3.
4.
5.
6.
7.

Score

Problem Solving

John has a monthly income of $5,500. What is his annual income?

Review Exercises

1. Find the perimeter.

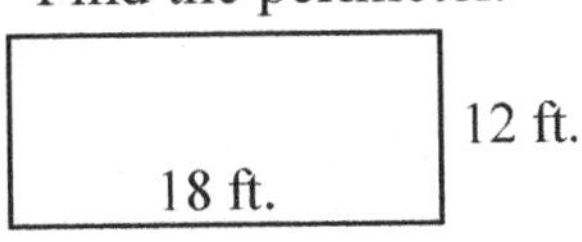

2. Find the perimeter.

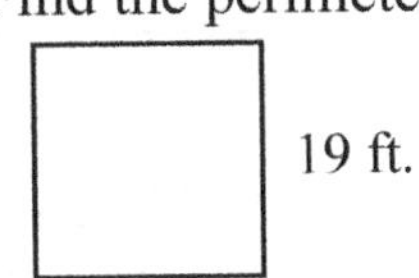

3. Find the area.

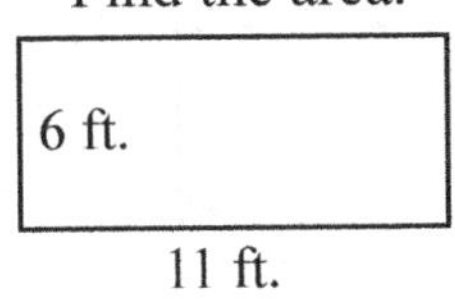

4. Find the area.

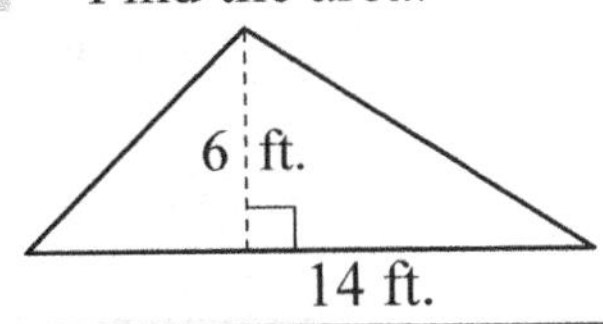

5. Find the area.

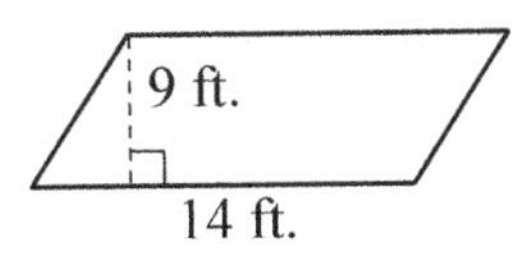

6. Find the circumference.

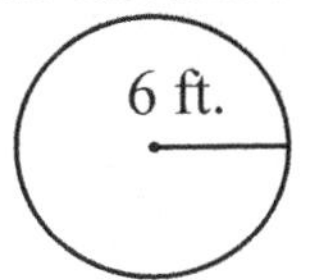

Helpful Hints

Triangle $A = \frac{b \times h}{2}$

Use what you have learned to solve the following problems.
* Remember, areas are expressed in square units.

Parallelogram $A = b \times h$

Find the following areas. Start with the formula. Substitute the values. Finally, solve the problem. If there is no diagram, draw a sketch.

S1.

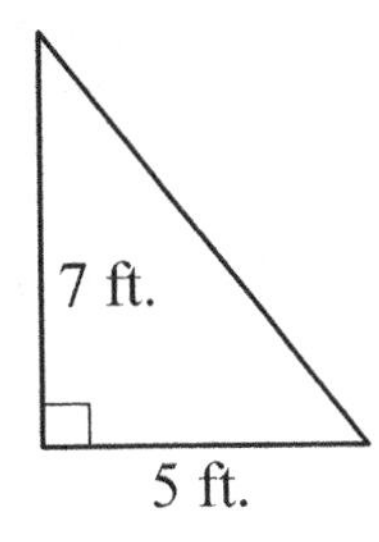

S2.

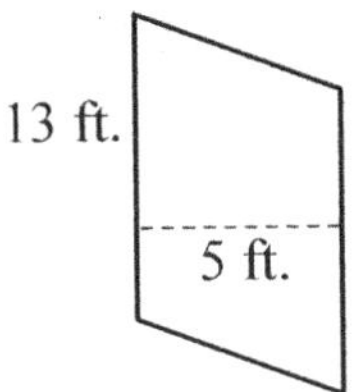

1.

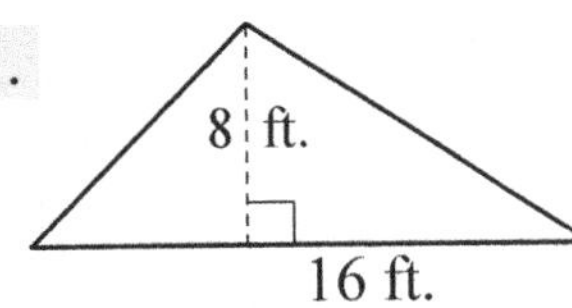

2.

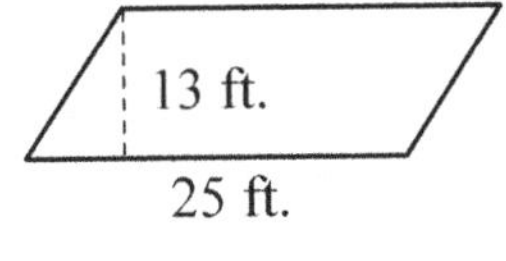

3.

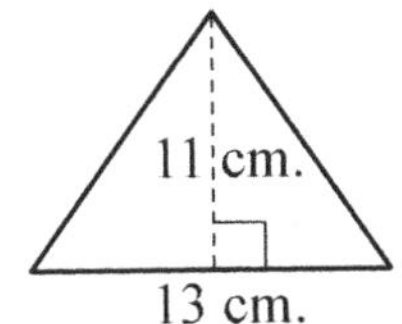

4. A parallelogram with a base of 22 ft. and a height of 12 ft.

5. 12 ft. 6 ft.

6. A triangle with a base of 20 cm. and a height of 12 cm.

7. The area of a parallelogram is 224 cm^2 and the base is 14 cm. What is the height?

1.

2.

3.

4.

5.

6.

7.

Score

Mary wants to replace the wall-to-wall carpet in her family room. If the room is 20 ft. by 16 ft., how many square feet of carpet will she need to cover the floor?

Review Exercises

For 1 - 6 find the areas.

1. 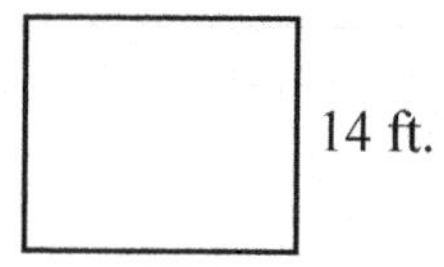

2. 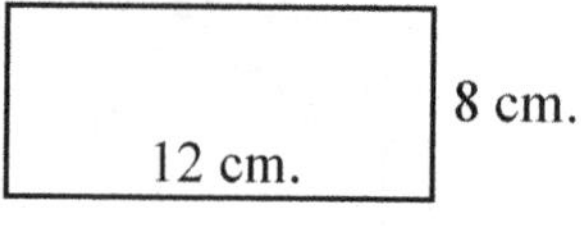

3. 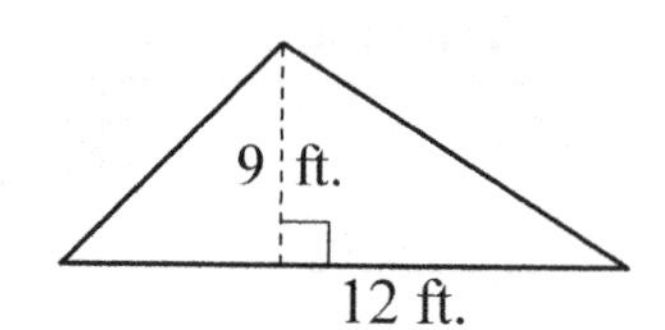

4. 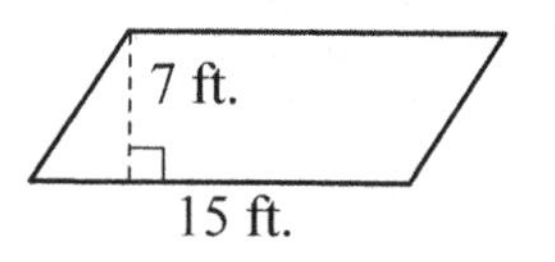

5. A triangle with b = 7 ft. and h = 15 ft.

6. A square with sides 8 ft.

Helpful Hints

Area of a trapezoid = $\frac{h(B+b)}{2}$

b

h

B

b = 6 ft.

h = 5 ft.

B = 9 ft.

Example: $A = \frac{h(B+b)}{2}$

$A = \frac{5(9+6)}{2}$

$A = \frac{5(15)}{2}$

$A = \frac{75}{2} = 2\overline{)75}$ → $37\frac{1}{2}$

Area = $37\frac{1}{2}$ **sq. ft.**

Find the following areas. If there is no figure, make a sketch. Write the formula. Substitute the values. Solve the problem.

S1. 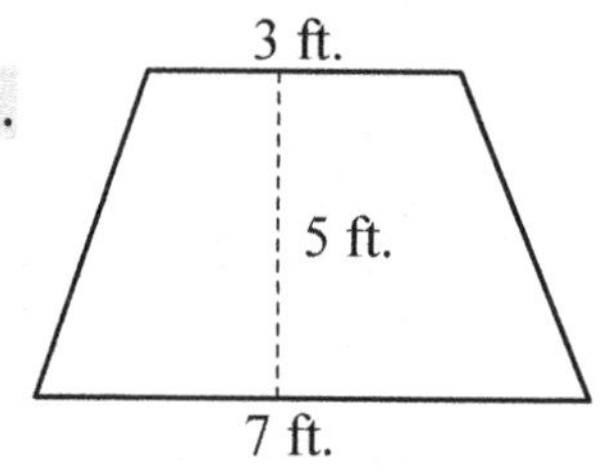

S2. 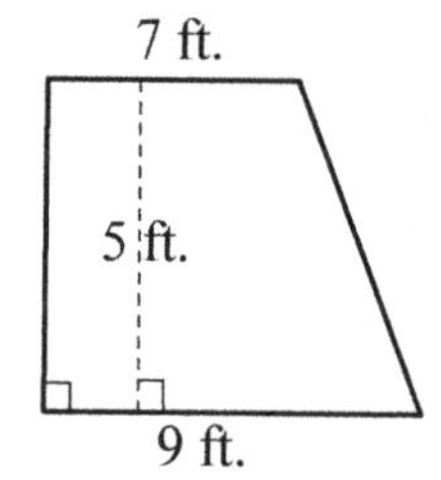

1. 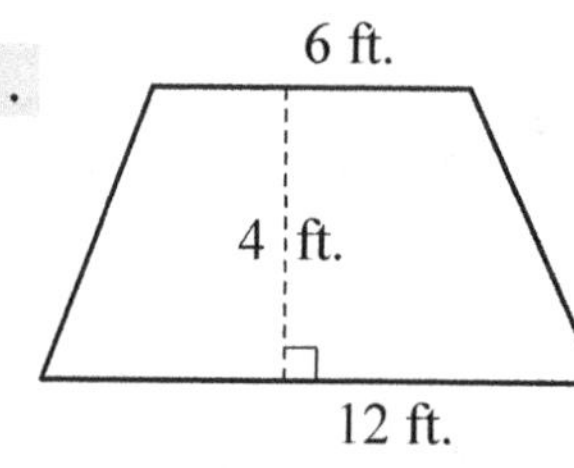

2. A Trapezoid with
B = 16 ft.
b = 6 ft.
h = 4 ft.

3. 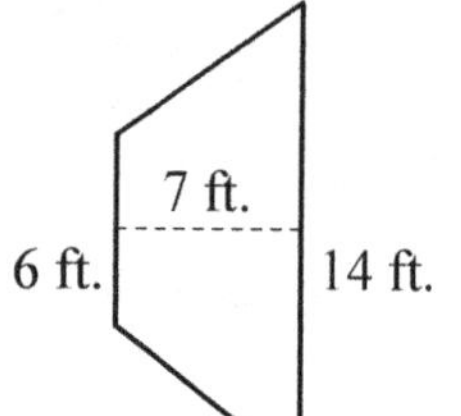

4. 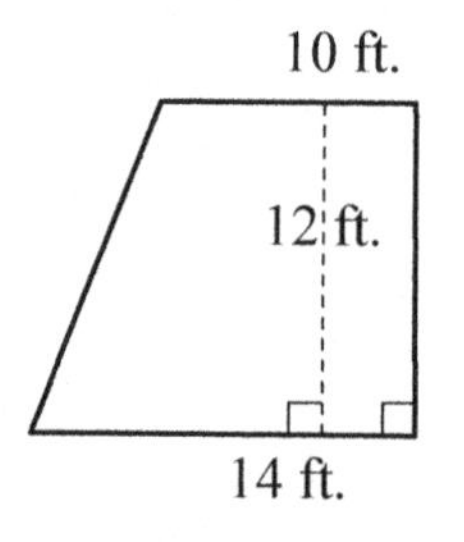

5. 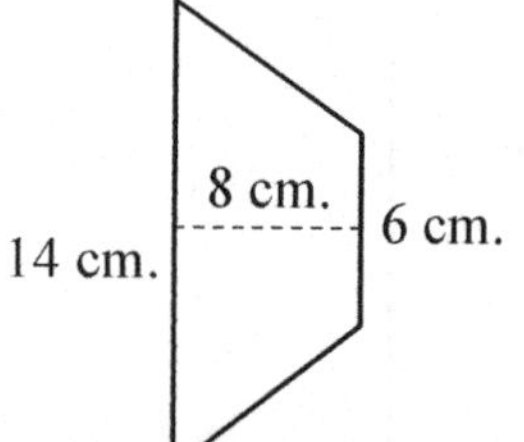

6. A Trapezoid with
B = 7 ft.
b = 4 ft.
h = 3 ft.

7. 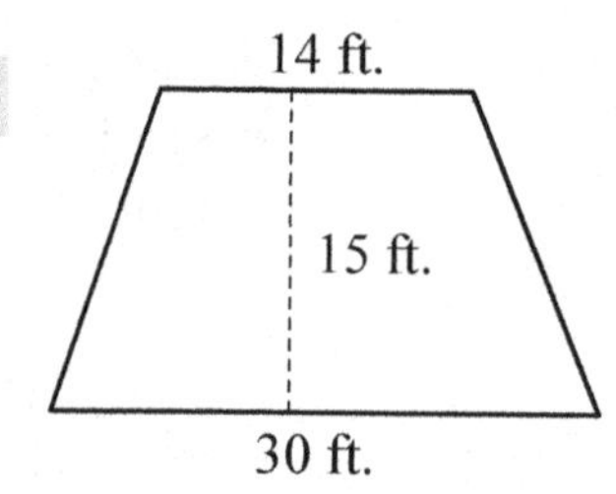

Problem Solving

Zoe is going to make a cloth banner in the shape of a triangle. If the base of the triangle is 24 inches and the height is 20 inches, how many square inches of cloth does she need?

1.

2.

3.

4.

5.

6.

7.

Score

Review Exercises

For 1 - 6 find the perimeter or circumference.

1.

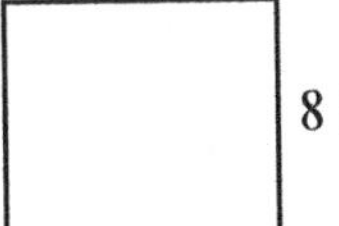

2.

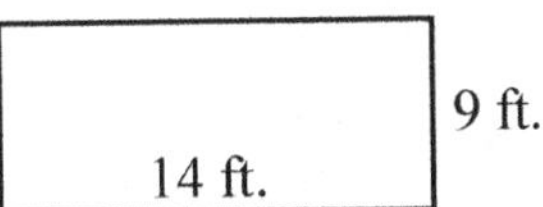

3. A regular pentagon with sides 17 ft.

4.

5. A regular decagon with sides 23 cm.

6.

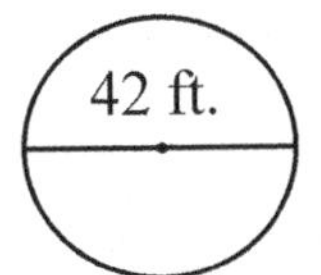

Helpful Hints

Trapezoid
$$A = \frac{h(B + b)}{2}$$

Use what you have learned to solve the following problems.
* Remember, areas are expressed in square units.

Find the following areas. If there is no figure, make a sketch. Write the formula. Substitute the values. Solve the problem.

S1.

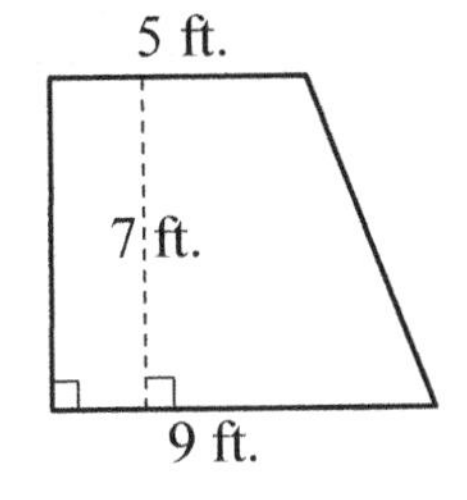

S2.

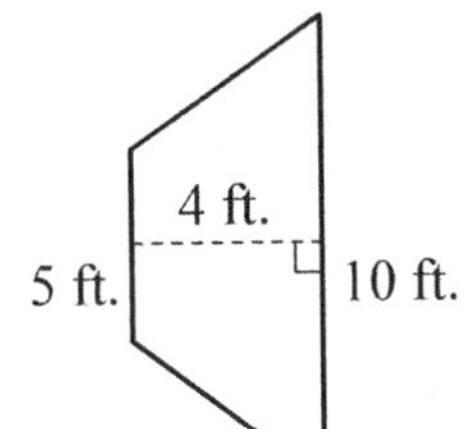

1. A Trapezoid with
B = 24 ft.
b = 10 ft.
h = 5 ft.

2.

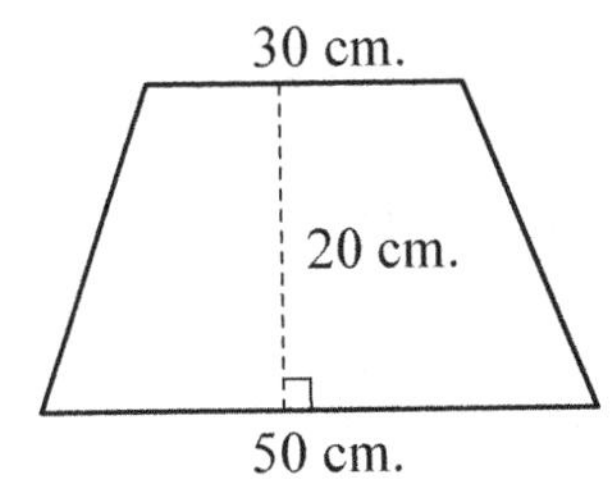

3. A Trapezoid with
B = 60 ft.
b = 50 ft.
h = 30 ft.

4.

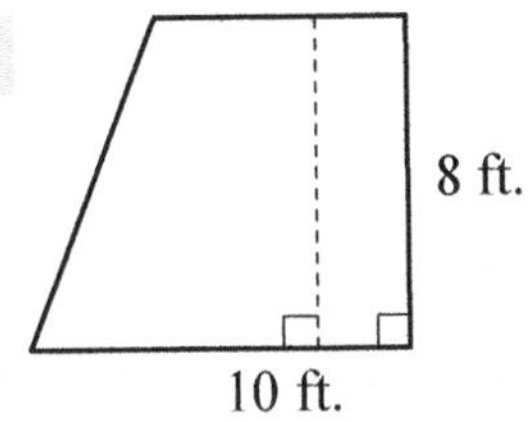

5.

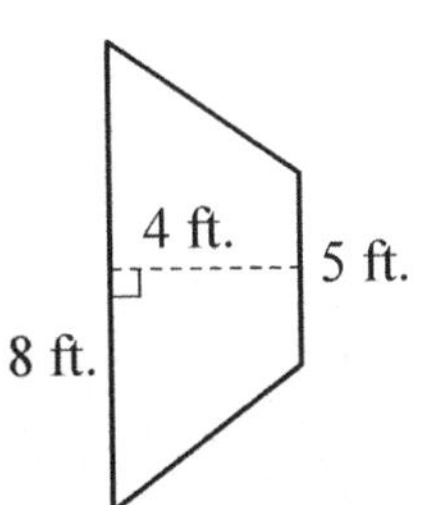

6.

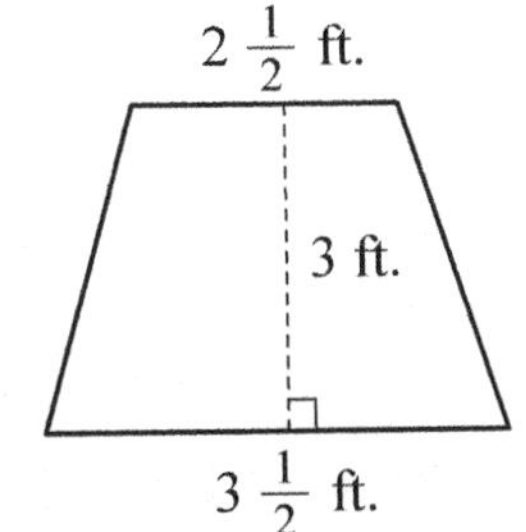

7.

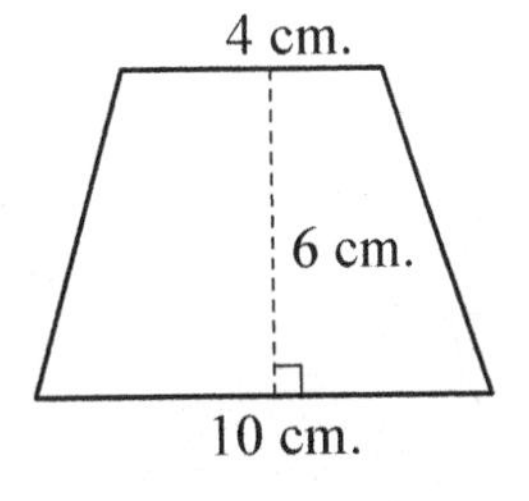

1.
2.
3.
4.
5.
6.
7.

Score

Problem Solving

A farmer has seven sacks of seed that weigh 250 pounds each. He has one barrel of apples that weighs 755 pounds. What is the total weight of the seed and apples?

Review Exercises

1. Find the area.

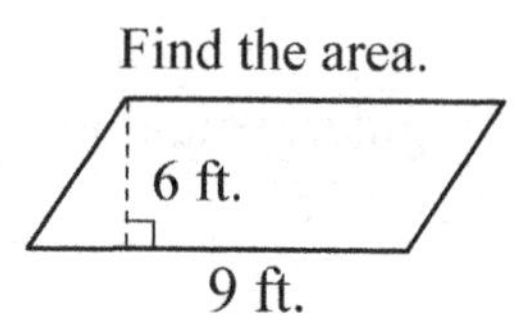

2. Find the area.

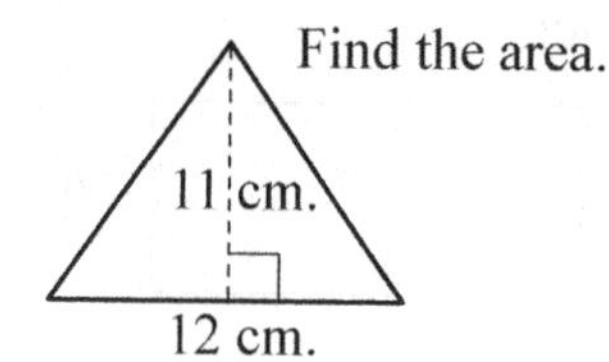

3. Find the perimeter.

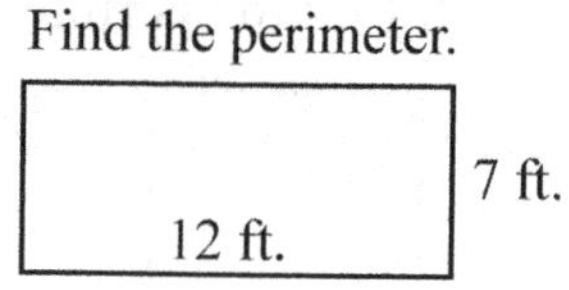

4. Find the perimeter.

14 ft.

5. Find the area.

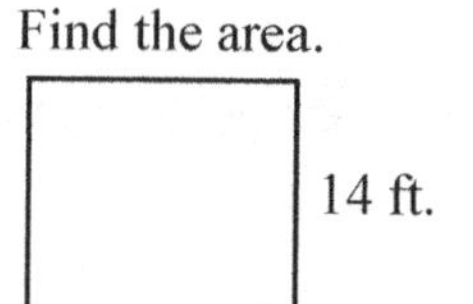

6. Find the area.

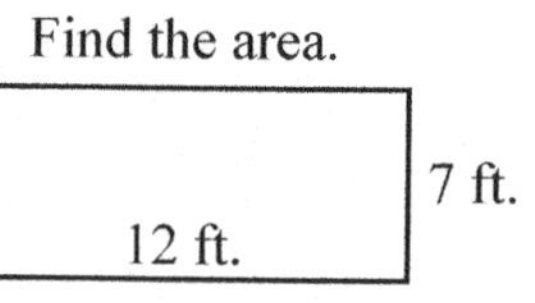

Helpful Hints

Area of a Circle = π × radius × radius
A = π × r × r = πr²
If the radius is divisible by 7, use $\pi = \frac{22}{7}$.

Examples:

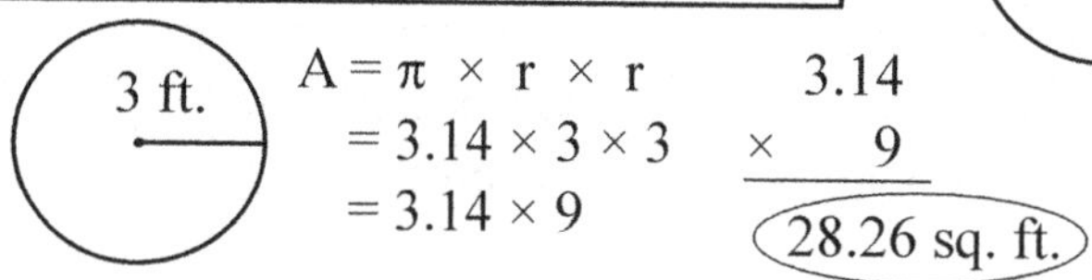

$A = \pi \times r \times r$
$= 3.14 \times 3 \times 3$
$= 3.14 \times 9$

3.14 × 9 = 28.26 sq. ft.

14 ft.

$A = \pi \times r \times r$
$= \frac{22}{7} \times \frac{7}{1} \times \frac{7}{1}$
$= 22 \times 7$

22 × 7 = 154 sq. ft.

Find the area of each circle. If there is no figure, draw a sketch. (sq. ft. = ft.^{2})
Write the formula. Substitute the values. Solve the problem.

S1.

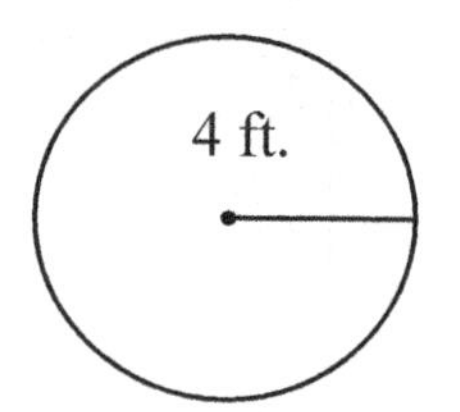

S2.

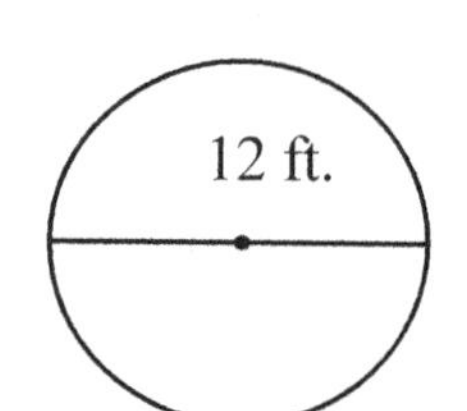

1.

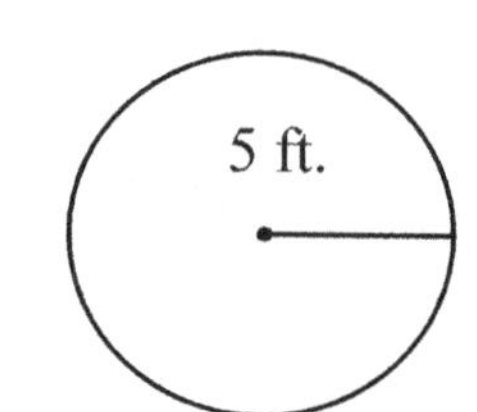

2.

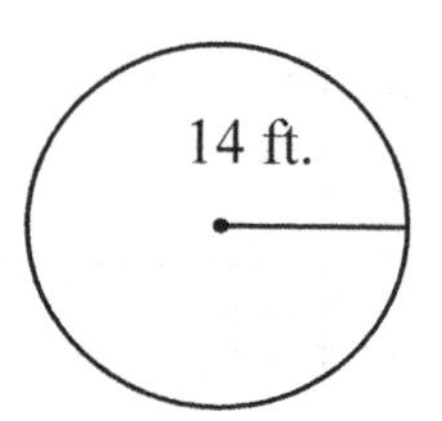

3.

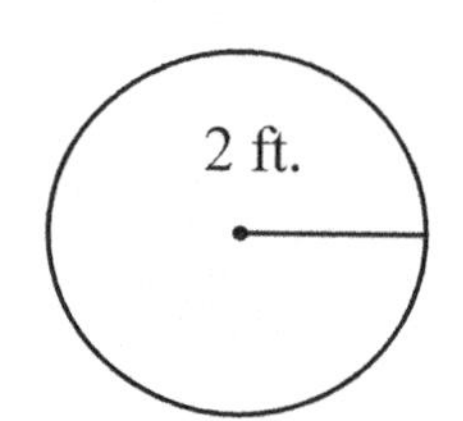

4.

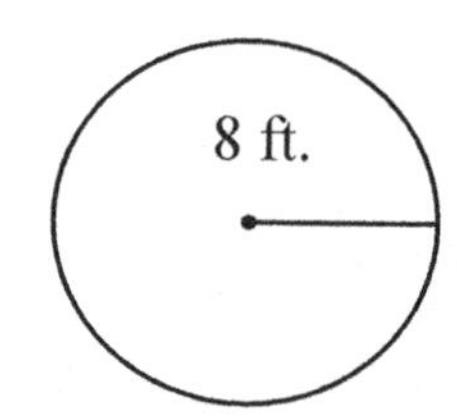

5.

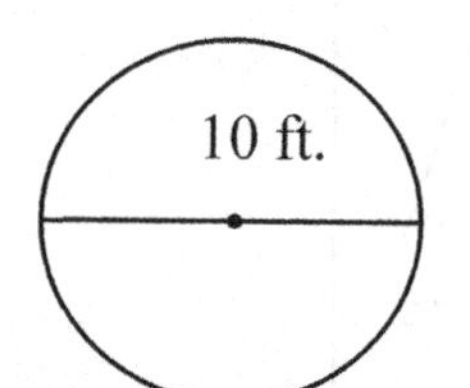

6. A circle with a radius of 6 ft.

7. A circle with a diameter of 14 ft.

1.
2.
3.
4.
5.
6.
7.

Score

Problem Solving

A city block is in the shape of a square. If the distance around the block is 1,280 ft., what are the lengths of each side of the block?

Review Exercises

1. Classify

2. Classify

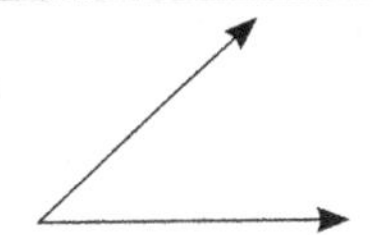

3. What is the supplement of 70°?

4. Classify

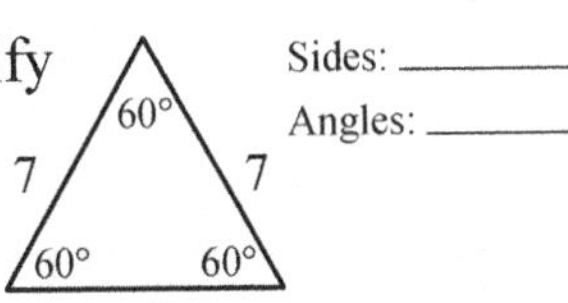

Sides: ________

Angles: ________

5. Find the circumference.

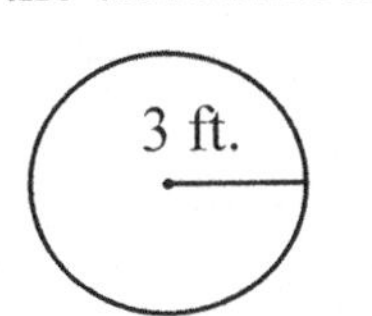

6. Find the area.

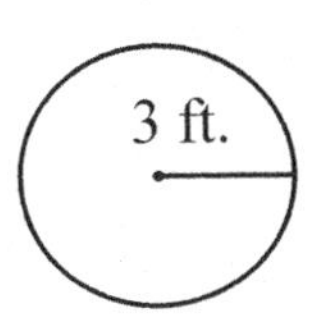

Helpful Hints

Use what you have learned to solve the following problems.

*Remember areas are expressed in square units.

*If the radius is divisible by 7, use $\pi = \frac{22}{7}$.

$A = \pi \times r \times r$

Find the area of each circle. If there is no figure, draw a sketch. Write the formula. Substitute the values. Solve the problem.

S1.

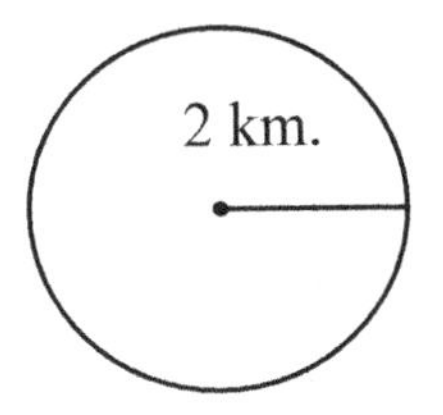

S2.

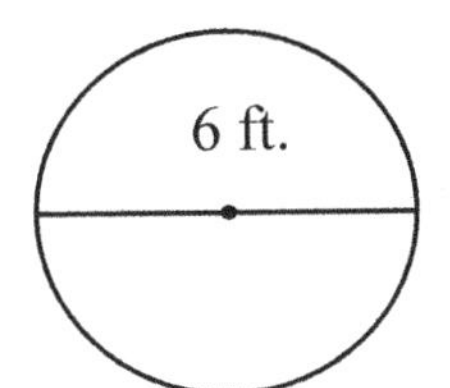

1.

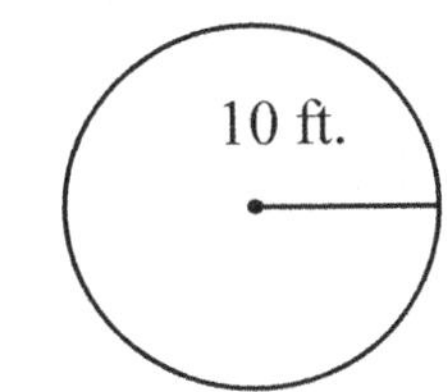

2.

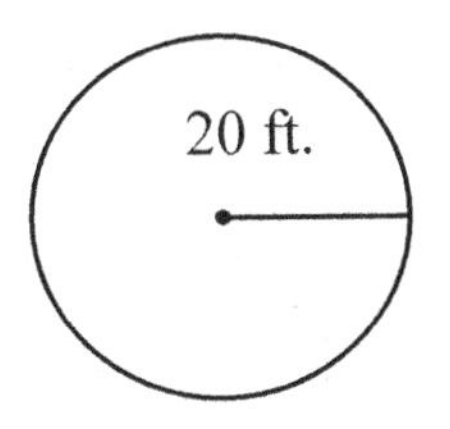

3. A circle with radius 14 ft.

4.

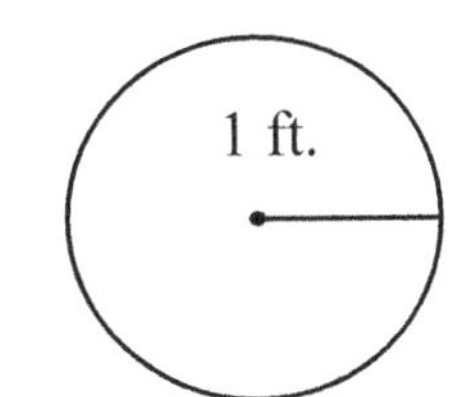

5. A circle with a diameter of 18 ft.

6. 42 ft.

7.

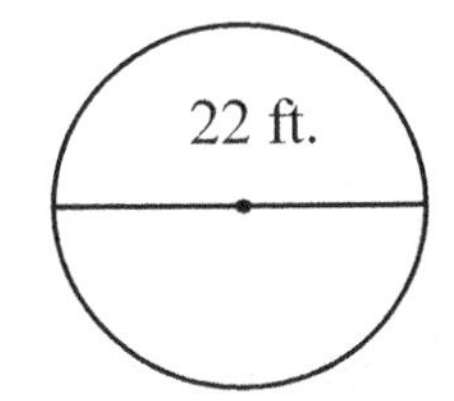

1.

2.

3.

4.

5.

6.

7.

Score

Problem Solving

A school has 280 boys and 320 girls. If the students are grouped into classes of 30 students each, how many classes are there?

Review Exercises

For 1-6, find the area of each figure.

1.

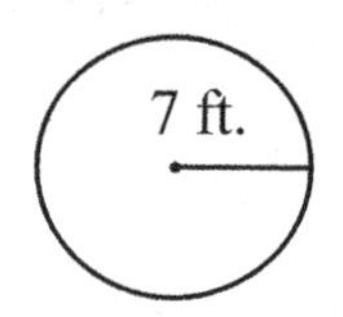

2.

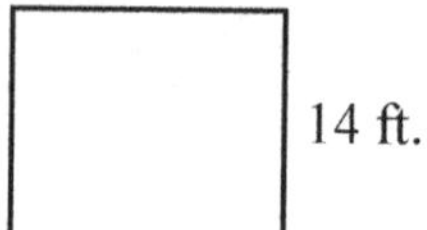

3.

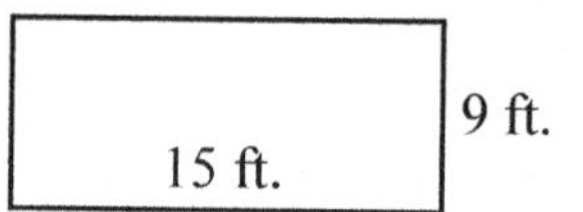

4.

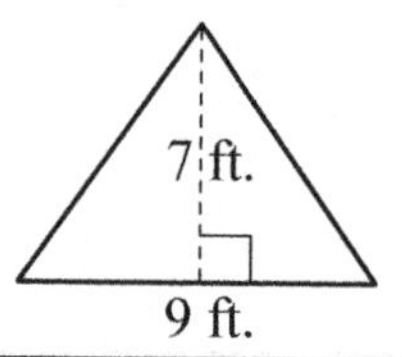

5.

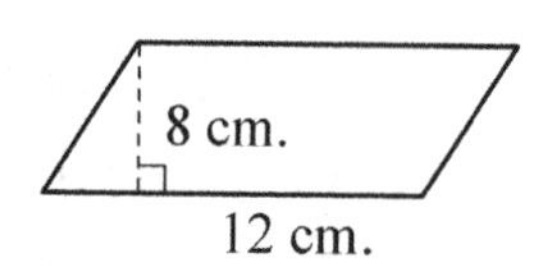

6.

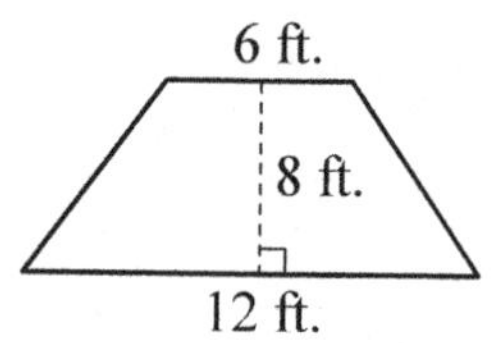

Helpful Hints

Remember these formulas. For areas: 1. Write the formula 2. Substitute values 3. Solve Problem

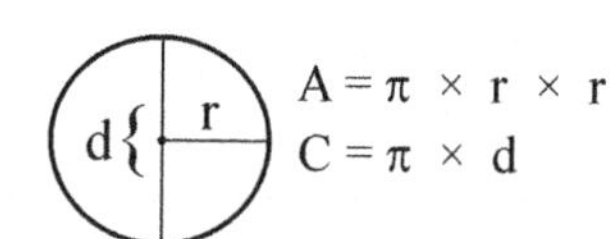

$A = \pi \times r \times r$
$C = \pi \times d$

$P = 4 \times s$
$A = s \times s$

h, b

$A = \dfrac{b \times h}{2}$

P = Sum of sides.

b, h, B

$A = \dfrac{h(B + b)}{2}$

P = Sum of sides.

l, w

$P = 2(l \times w)$
$A = l \times w$

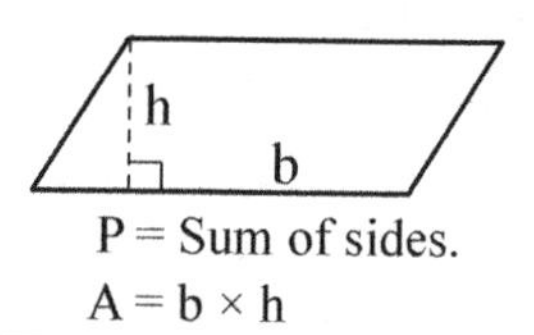

P = Sum of sides.
$A = b \times h$

Find the perimeter or circumference then find the area.

S1.

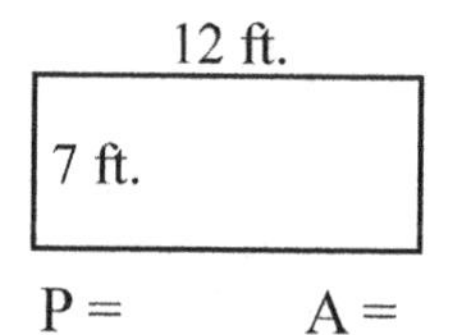

P = A =

S2.

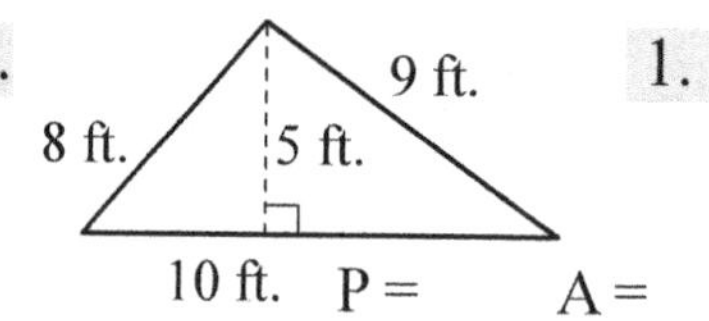

P = A =

1. P = A =

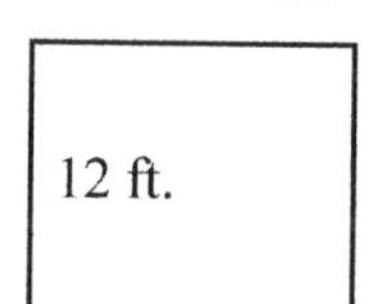

2.

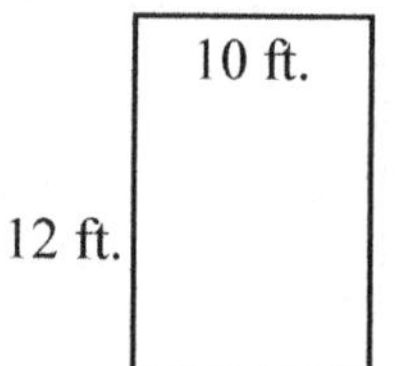

P =
A =

3.

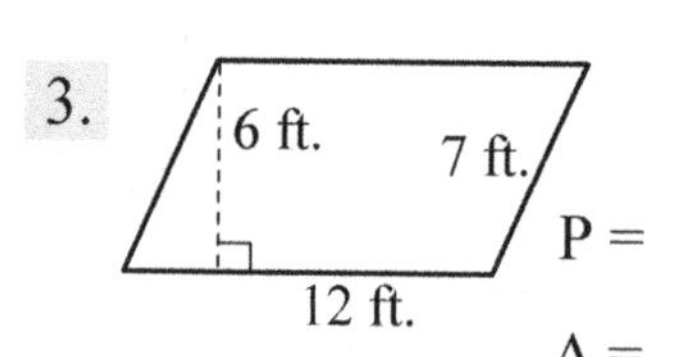

P =
A =

4.

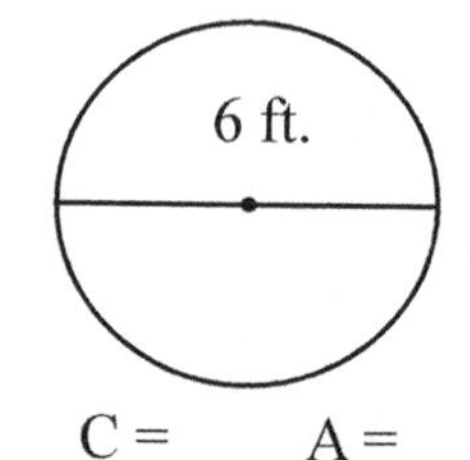

C = A =

5.

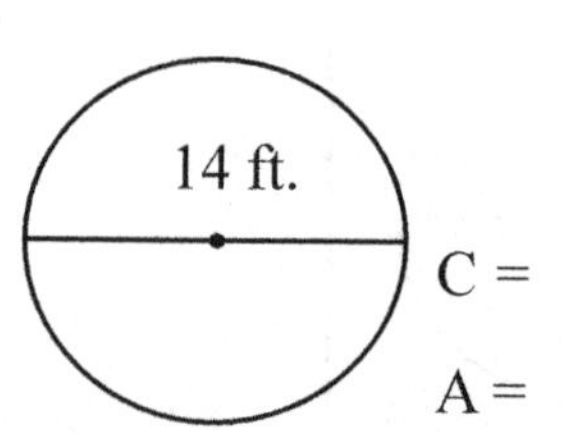

C =
A =

6.

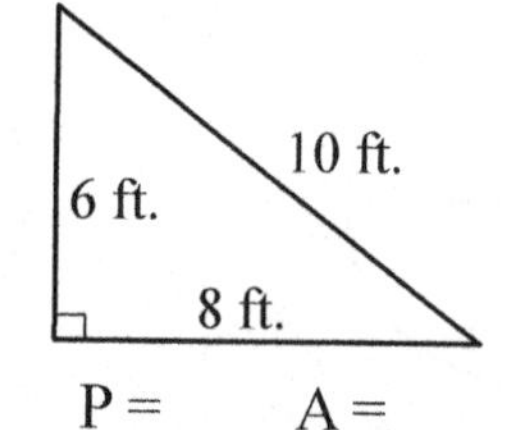

P = A =

7.

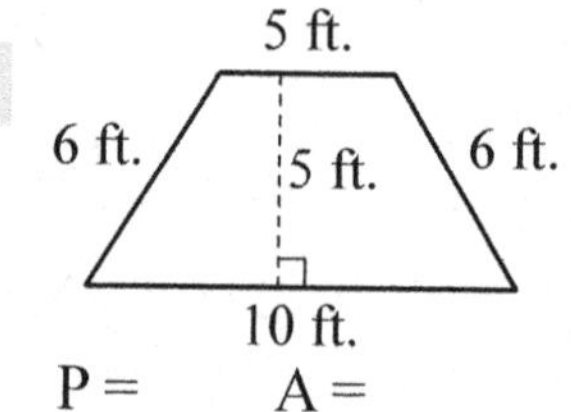

P = A =

1.
2.
3.
4.
5.
6.
7.
Score

Problem Solving

Paul worked nine hours on Monday and eight hours each of the next three days. If he earned 12 dollars per hour, what was his pay?

Review Exercises

For 1-6, find the perimeter of each figure.

1. 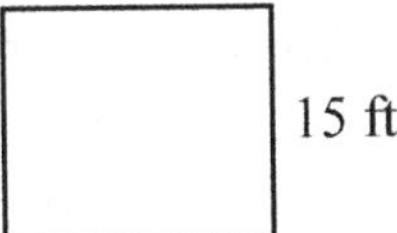

2. 24 ft. 18 ft.

3. 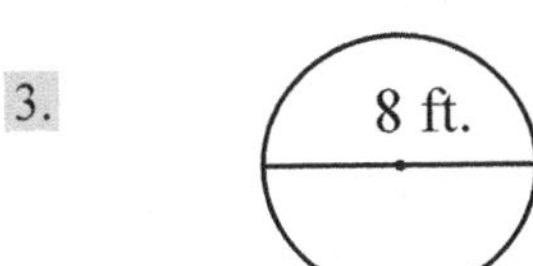

4. 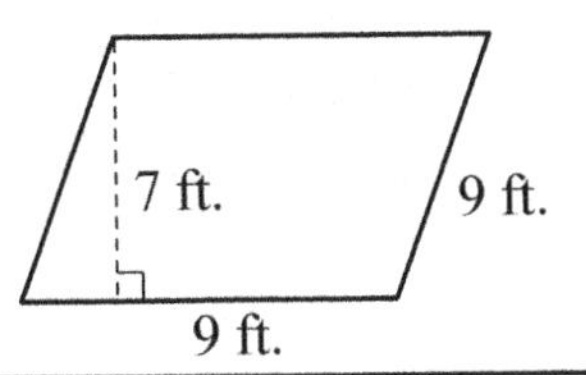

5. 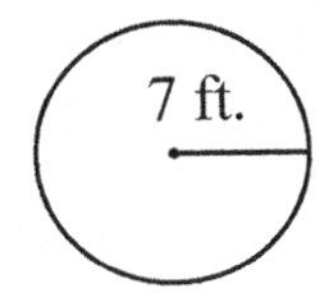

6. 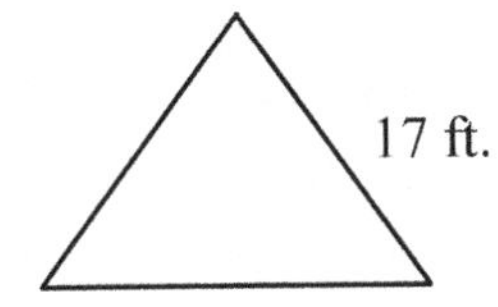

Helpful Hints

Use what you have learned to solve the following problems.
* Perimeter or circumference are distances around a figure.
* Areas are expressed in square units.

Find the perimeter or circumference then find the area.

S1. 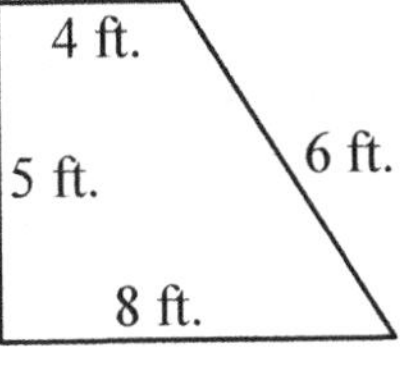

P = A =

S2. 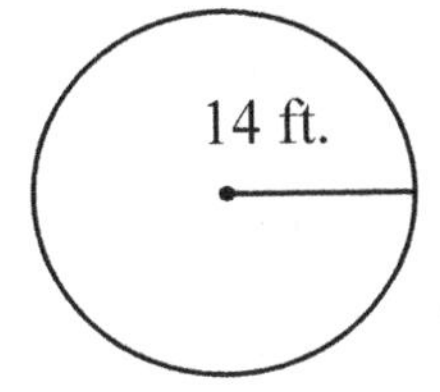

C = A =

1. 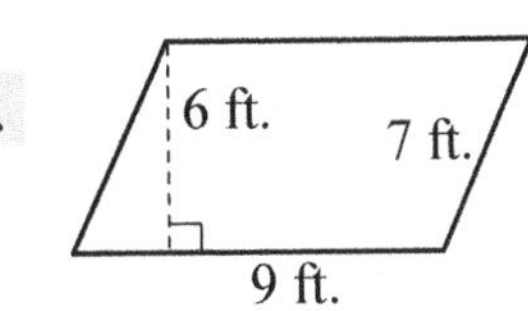

P = A =

2. 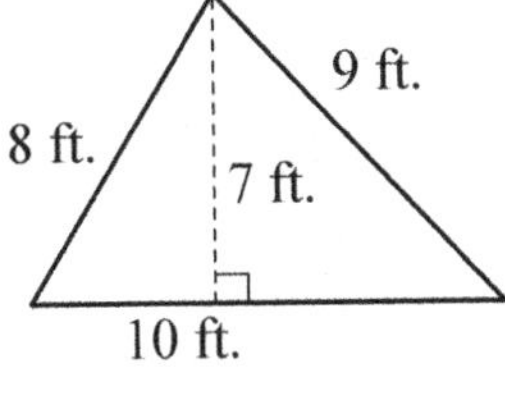

P = A =

3. 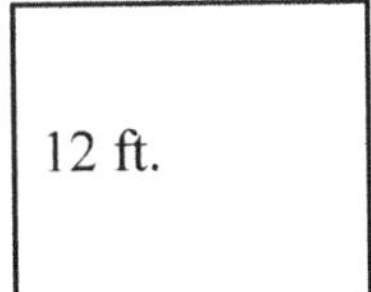
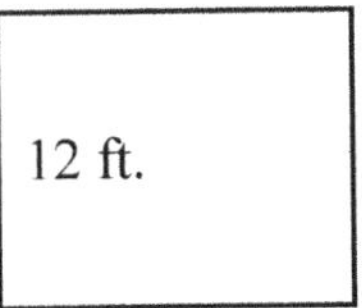

P = A =

4. 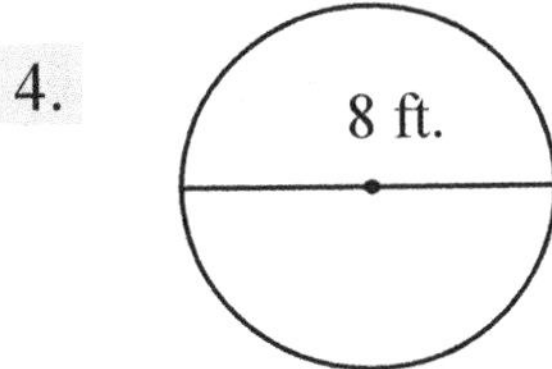

C = A =

5. 29 ft. 14 ft.

P = A =

6. 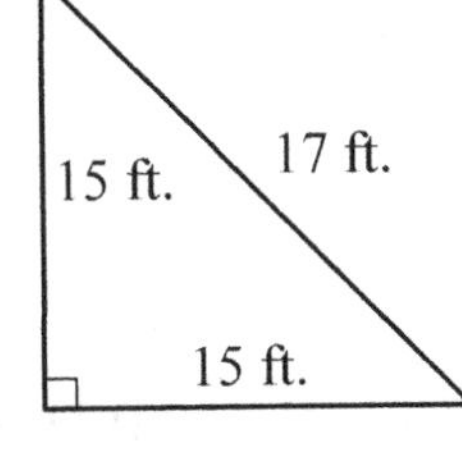

P = A =

7. 6 ft. 7 ft. 6 ft. 7 ft. 8 ft.

P = A =

1.

2.

3.

4.

5.

6.

7.

Score

Problem Solving

Five pounds of dog food costs $6.75.
What is the price per pound?

Review Exercises

For 1 - 6 draw a sketch and find the area.

1. **Square:** Sides, 16 ft.

2. **Triangle:** Base, 18 ft. Height, 8 ft.

3. **Parallelogram:** Base, 22 ft. Height, 16 ft.

4. **Trapezoid:** B, 22 ft. b, 20 ft. h, 6 ft.

5. **Rectangle:** Length, 16 ft. Width, 18 ft.

6. **Circle:** Radius, 5 ft.

Helpful Hints

Use what you have learned to solve the following problems.

1. Read the problem carefully to understand what is being asked.
2. Draw a sketch.
3. Write the formula. Substitute the values. Solve the problem. Label the answer with a word or short phrase.

S1. The Smith's kitchen floor is 12 ft. by 14 ft. What is the perimeter of the floor?

S2. Phil's garden is 12 ft. by 10 ft. If each bag of fertilizer will cover 6 sq. ft., how many bags of fertilizer will he need to care of his garden?

1. Anna wants to build a fence around her yard. If the rectangular yard is 28 ft. by 20 ft., how many feet long will the fence be?

2. Eddie wants to put crown molding around the ceiling of his living room. The room is rectangular and is 24 ft. by 18 ft. If molding comes in 6 foot sections, how many sections of molding must he buy?

3. Find the circumference of a circular flower bed with radius 8 ft.

4. Sue wants to paint a rectangular wall that is 15 ft. by 9 ft. A can of paint covers 45 sq. ft. How many cans must she buy?

5. The perimeter of a square is 624 ft. What is the length of its sides?

6. A city is in the shape of a rectangle. Its area is 192 sq. miles and its width is 12 miles. What is the city's length?

7. A walking path is in the shape of an equilateral triangle. If Dan has walked the first two sides and covered 1260 meters, what is the length of each side of the walking path?

1.
2.
3.
4.
5.
6.
7.
Score

Problem Solving

Mr. Jenkins earned \$3,600 and deposited $\frac{1}{3}$ of it into his savings account. How much did he deposit into his savings account?

Review Exercises

1. What angle compliments 35°?

2. What angle supplements 78°?

3. An equilateral triangle has a perimeter of 108 ft. What is the length of each side?

4. The perimeter of the square is 940 ft. What is the length of $\overline{AB}$?

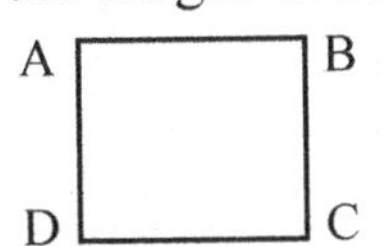

5. Find the hypotenuse. (Round to the nearest whole number.)

16 ft.
18 ft.

6. 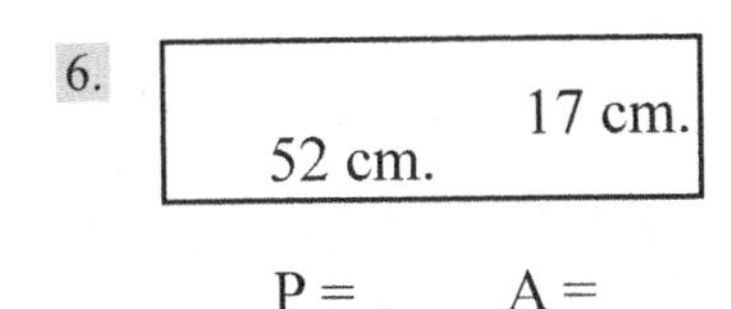

P = A =

Helpful Hints

Use what you have learned to solve the following problems.
1. Read the problem carefully to understand what is being asked.
2. Draw a sketch.
3. Write the formula. Substitute the values. Solve the problem. Label the answer with a word or short phrase.

S1. A park is circular shaped with a radius of 2 miles. If you walk halfway around the park, how far will you walk?

S2. Bill wants to carpet a rectangular room that is 12 ft. by 10 ft. If carpet costs $12 per square foot, what will be the cost of the carpet?

1. The radius of a circular race track is 7 miles. If you drive three laps around the track, how many miles will be traveled?

2. A developer has a lot that is 1,200 feet by 800 feet. If he wants to divide it into 24 equally-sized sections, how many square feet will be in each section?

3. Sara wants to trim a square window with sides 24 inches. How many inches of trim does she need?

4. Jill wants to sew a circular emblem that has a diameter of 6 inches. How many square inches of cloth will she need?

5. A stop sign his in the shape of a regular octagon with sides 18 inches. What is the perimeter of the sign?

6. Find the area of a trapezoid if the sum of the bases is 14 feet and the height is eight feet.

7. Find the area of a triangle if the base is 12 feet and the height is twice the length of the base.

1.
2.
3.
4.
5.
6.
7.
Score

Problem Solving

It takes Mr. Jones 12 minutes to commute to work. It only takes 10 minutes to drive home. How many minutes does it take him to drive to and from work in a 5-day work week?

Review Exercises

1. Find the circumference.

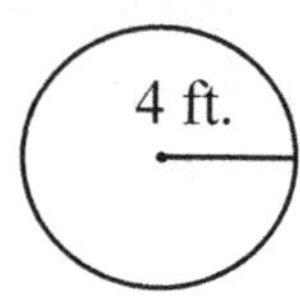

2. Find the area.

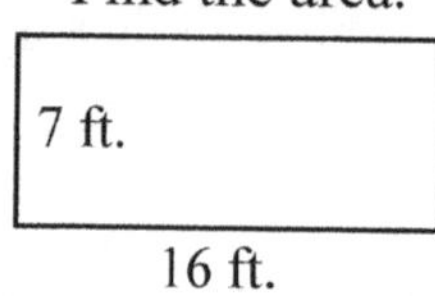

3. Find the perimeter.

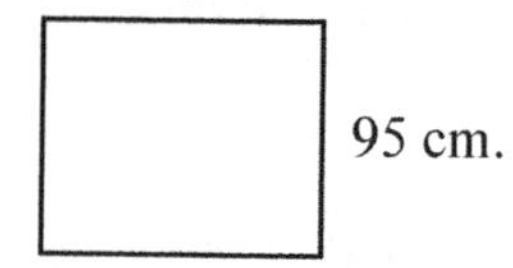

4. Find the area.

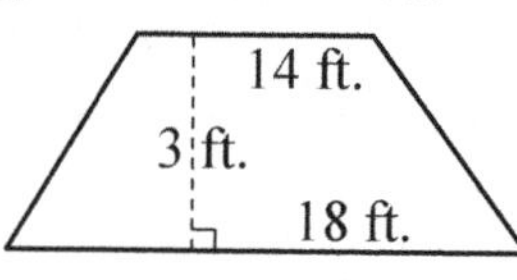

5. Find the circumference.

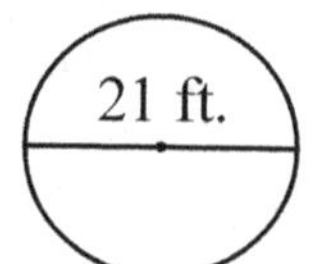

6. Find the area.

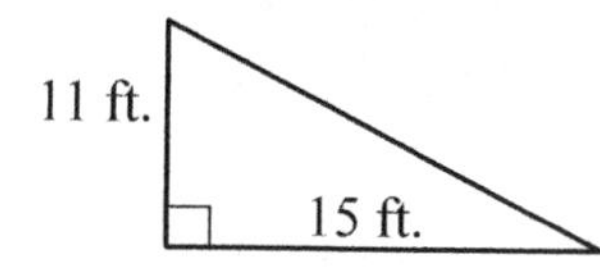

Helpful Hints

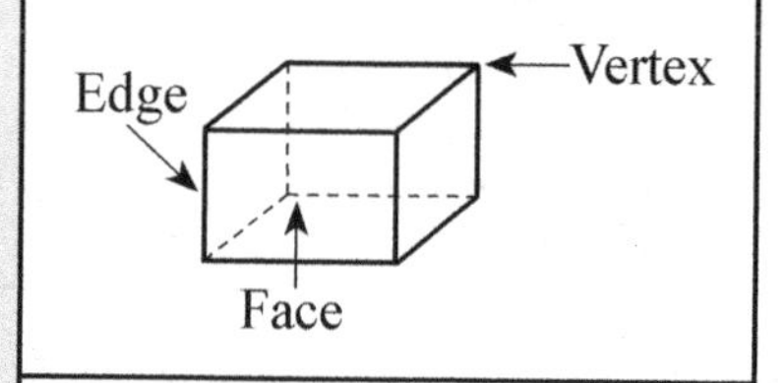

Triangular Prism

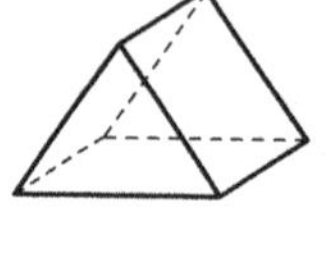

Triangular Pyramid

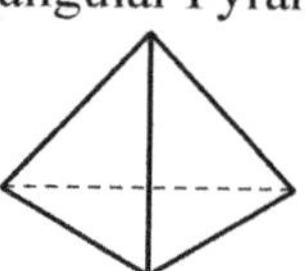

Square Pyramid

Cube

Rectangular Prism

Sphere

Cone

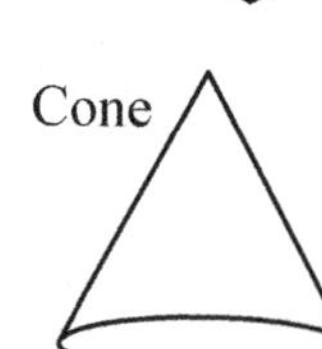

Cones and cylinders do not have straight edges.

Cylinder

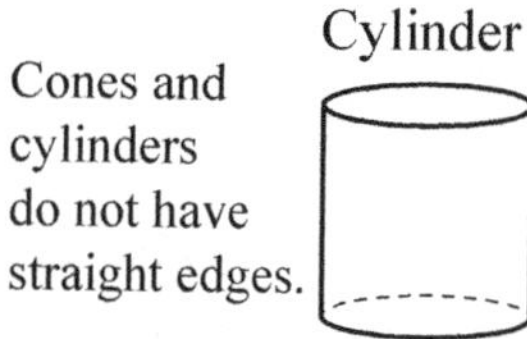

Identify the shape and the number of each part for each solid figure.

S1.

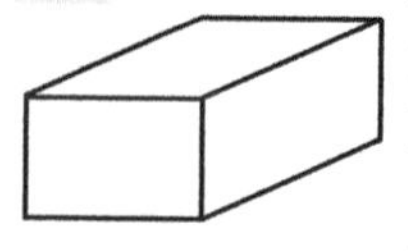

Name:________
Faces:________
Edges:________
Vertices:________

S2.

Name:________
Faces:________
Edges:________
Vertices:________

1.

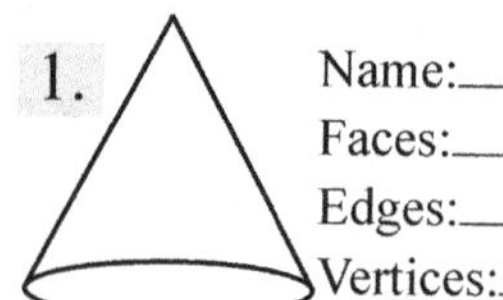

Name:________
Faces:________
Edges:________
Vertices:________

2.

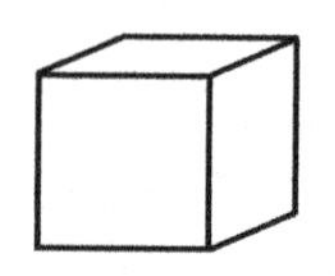

Name:________
Faces:________
Edges:________
Vertices:________

3.

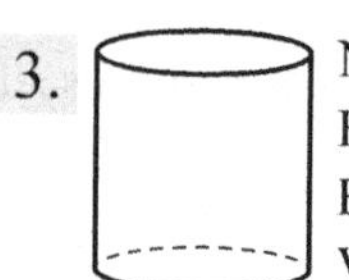

Name:________
Faces:________
Edges:________
Vertices:________

4.

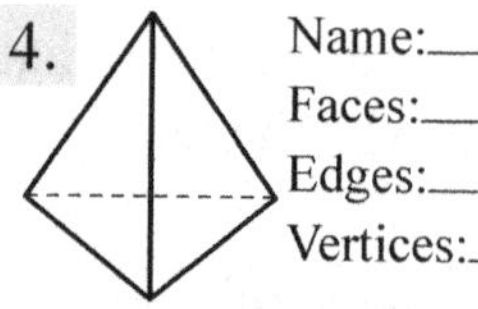

Name:________
Faces:________
Edges:________
Vertices:________

5.

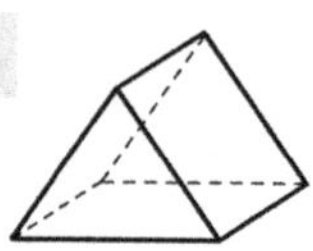

Name:________
Faces:________
Edges:________
Vertices:________

6. Name:________

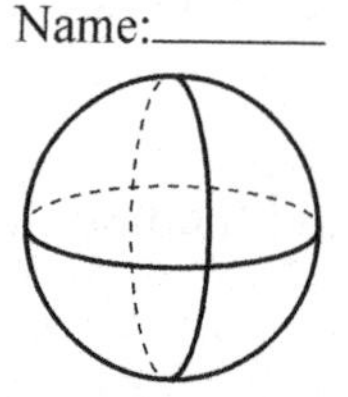

7. How many more faces does a cube have than a triangular prism?

1.
2.
3.
4.
5.
6.
7.
Score

Problem Solving

Cookies cost $.32 each. How much will two dozen cost?

Review Exercises

1. Name a solid shape with five faces.
2. Name a solid shape with six faces.
3. Which solid shape has no faces?
4. Name two solid shapes with 12 edges.
5. Name a solid shape with five vertices.
6. Name the solid shape with one or two faces.

Helpful Hints

Use what you have learned about solid shapes to answer the following questions.

For each solid shape list four real life objects that have that shape.

S1. Sphere

S2. Triangular Prism

1. Cube

2. Cone

3. Cylinder

4. Rectangular Prism

5. Square Pyramid

6. Triangular Pyramid

7. Objects made up of a combination of solid shapes.

1.
2.
3.
4.
5.
6.
7.
8.
9.
10.
Score

Problem Solving

If a city has an average monthly rainfall of 15 inches, what is its average annual rainfall?

Review Exercises

1. Find the complement of 18°.
2. Find the supplement of 114°.
3. Find the area of a circle with a radius of five feet.
4. Find the circumference of a circle with a radius of five feet.
5. Find the perimeter of a regular hexagon with sides of 15 ft.
6. Find the area of a square with sides of 16 ft.

Helpful Hints

The **surface area** of a **cube** or **rectangular prism** is the total area of all the sides.

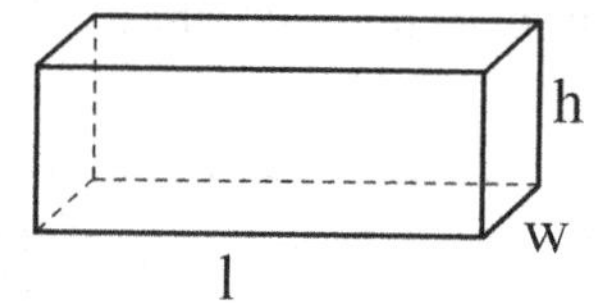

Example:

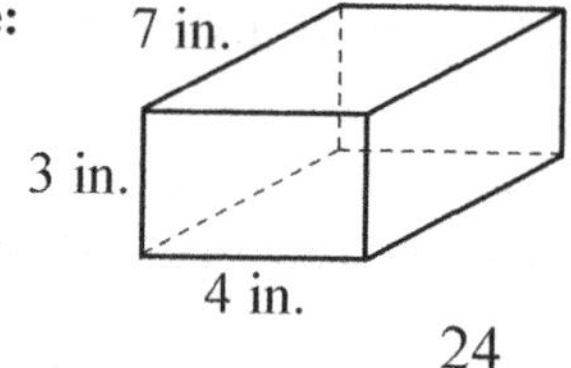

1. Area of front and back. $(4 \times 3) \times 2 = 24$
2. Area of top and bottom. $(7 \times 4) \times 2 = 56$
3. Area of both sides. $(3 \times 7) \times 2 = 42$

Simply find the area of each side and add them together.

$$24 + 56 + 42 = 122 \text{ in.}^2$$

Find the surface area of each cube or rectangular prism.

S1.

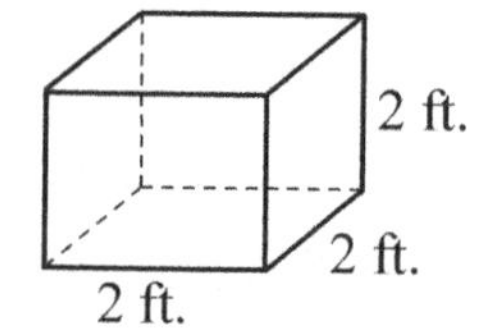

S2.

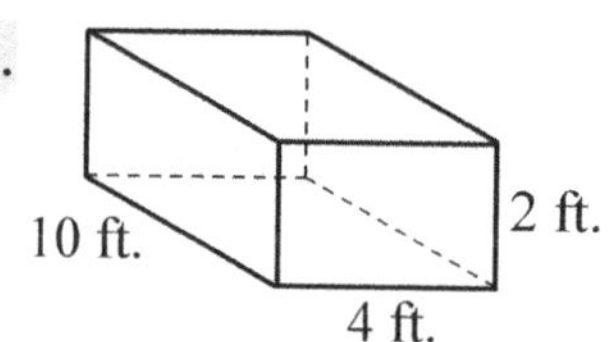

1.

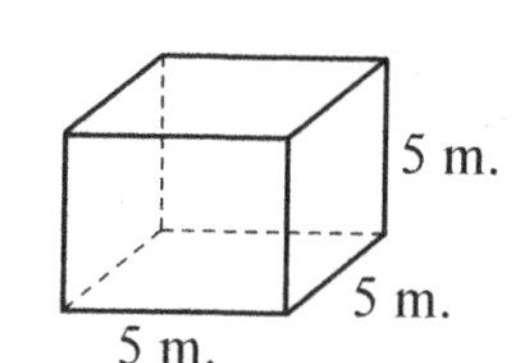

2.

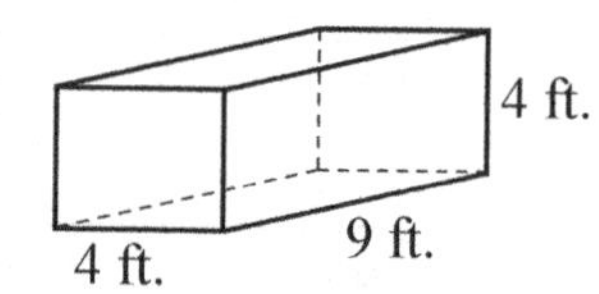

3.

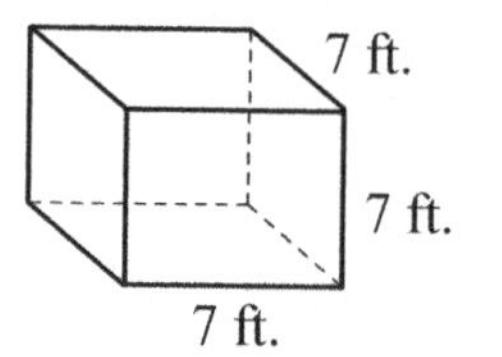

4.

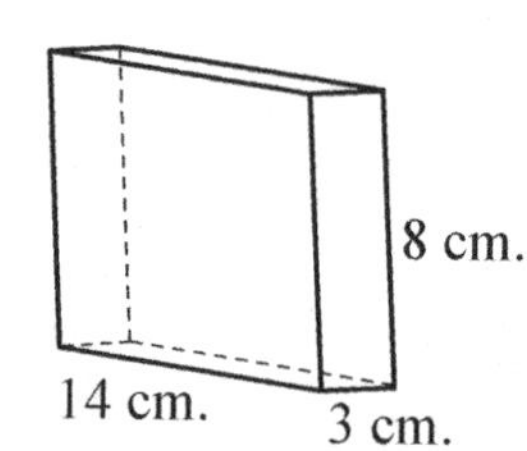

5.

9 cm.
6 cm.
5 cm.

6.

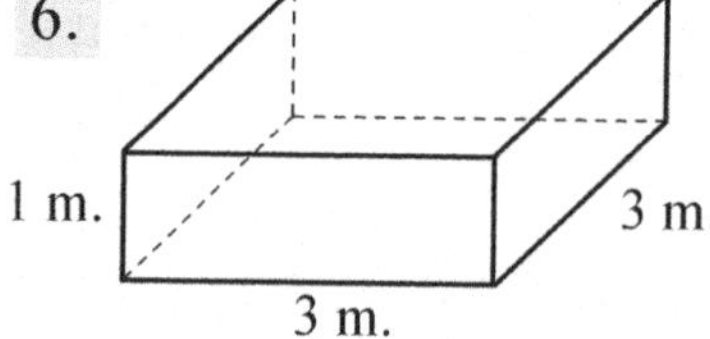

7.

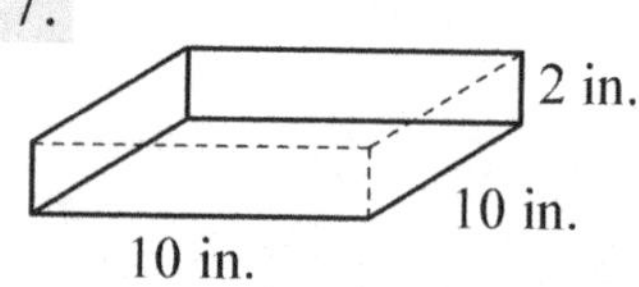

1. ____
2. ____
3. ____
4. ____
5. ____
6. ____
7. ____

Score

Problem Solving

A ranch in Texas is 12 miles long and 7 miles wide. What is the area of the ranch?

Review Exercises

For 1 - 6 find the area of each figure.

1. 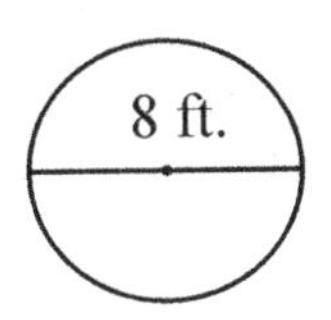

2. 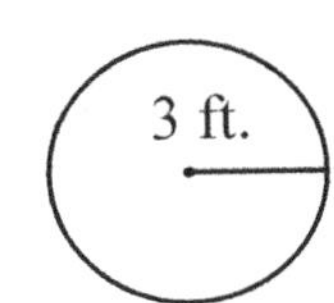

3. 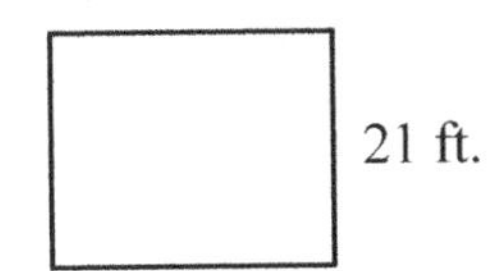

4. 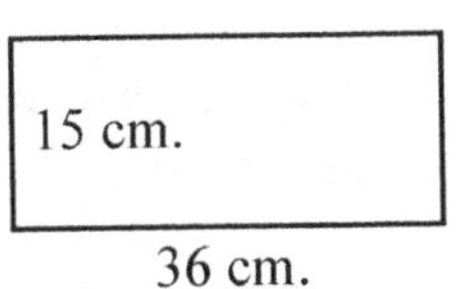

5. 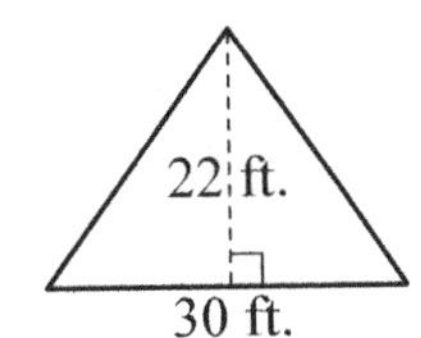

6. 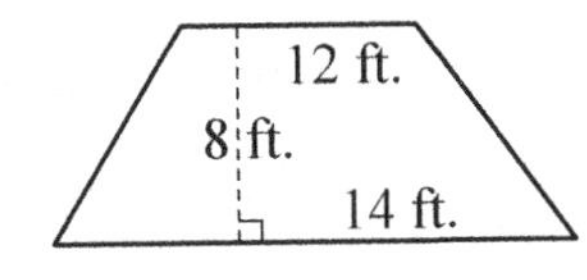

Helpful Hints

Use what you have learned to solve the following problems.
* Surface area is expressed in square units.

Find the surface area of each cube and rectangular prism.

S1. 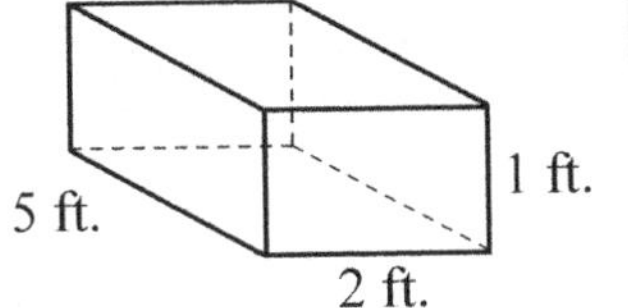

S2. 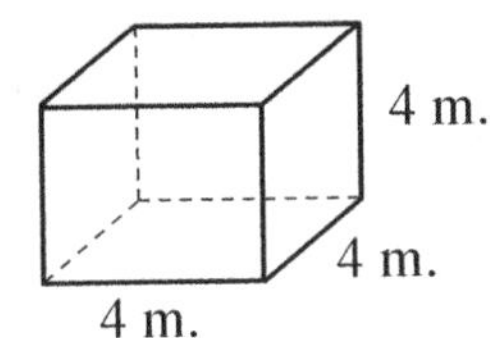

1. 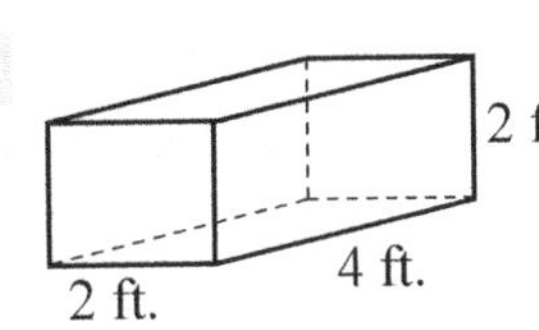

2. 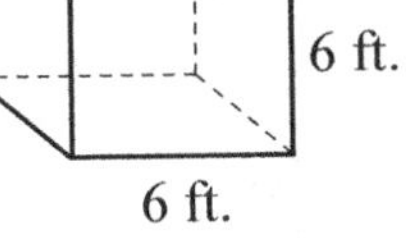

3. 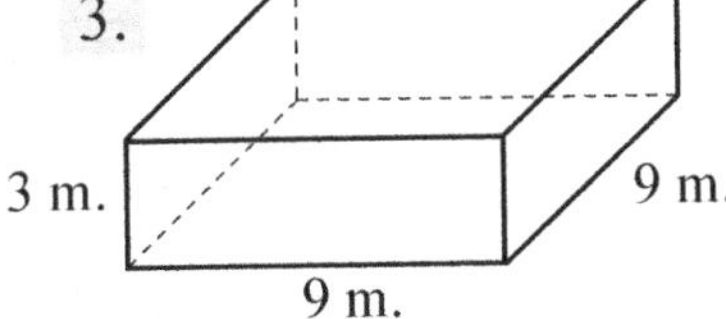

4. 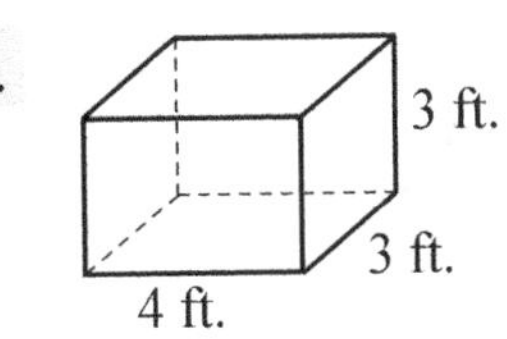

5. 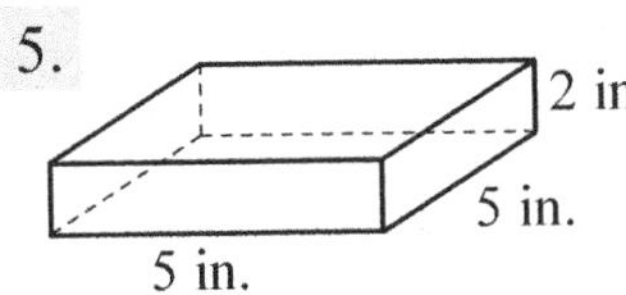

6. 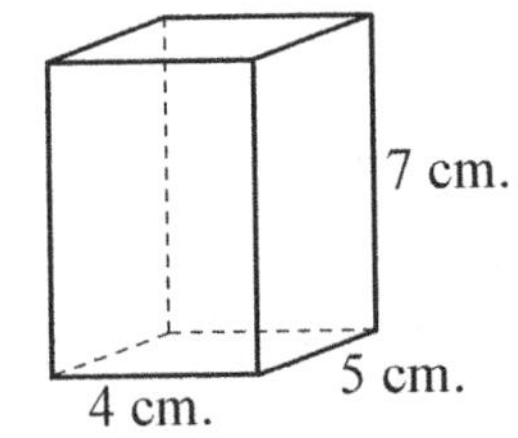

7. 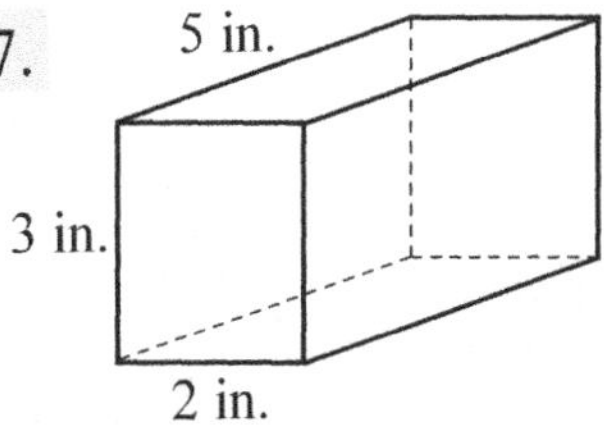

1.

2.

3.

4.

5.

6.

7.

Score

Problem Solving

A school wants to put a new wood floor on a sports court that is 40 ft. by 24 ft. If flooring costs $ 20 per square foot, how much will the new floor cost?

Review Exercises

For 1 - 6 find the perimeter or circumference.

1. A square with sides of 29 ft.
2. A rectangle with length 16 ft. and width 7 ft.
3. A circle with a diameter of 14 ft.
4. A circle with a radius of 3 ft.
5. A regular pentagon with sides of 14 ft.
6. An equilateral triangle with sides of 38 ft.

Helpful Hints

The **volume** of a solid is the number of cubic units that can be contained in the solid. Volume is expressed in cubic units such as in.^{3} or ft.^{3}

h w l

Example:

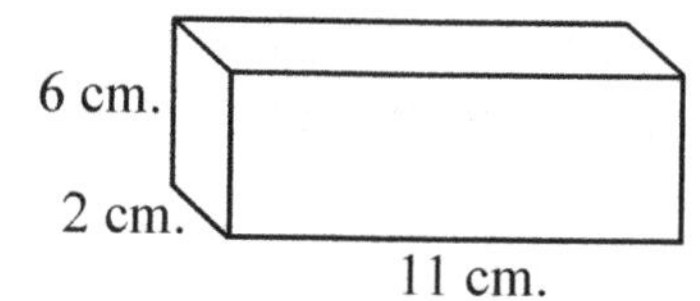

To find the volume of a cube or rectangular prism, multiply the length × width × height.

$V = l \times w \times h$

$V = 11 \times 2 \times 6$

$V = \boxed{132 \text{ cm.}^3}$

Find the volume of each cube and rectangular prism.

S1.

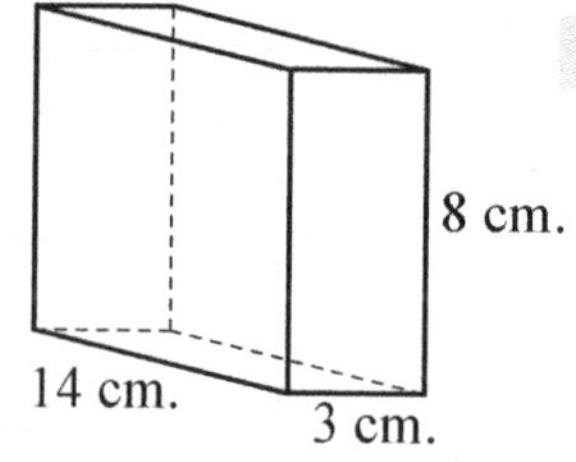

S2.

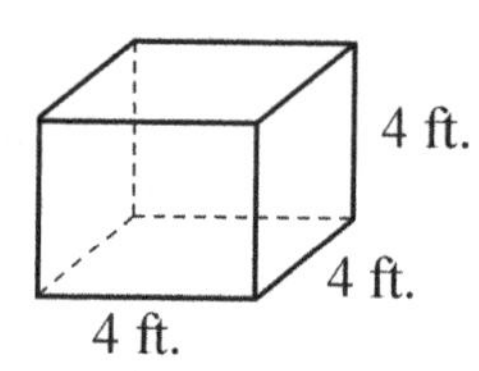

1.

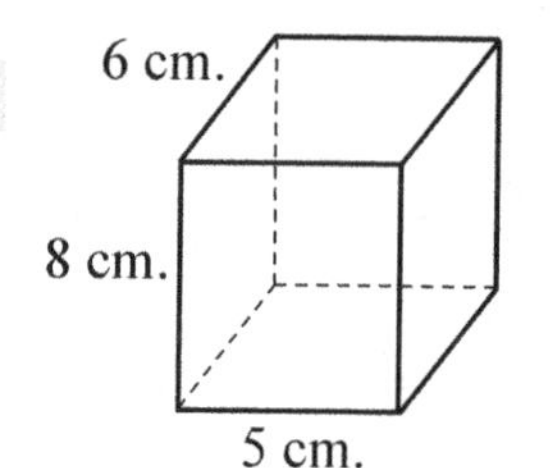

2.

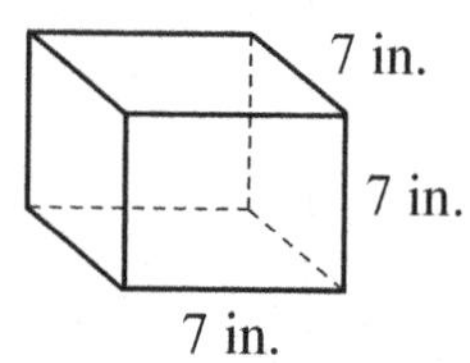

3.

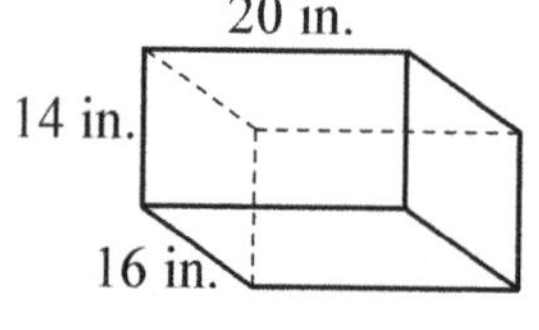

4.

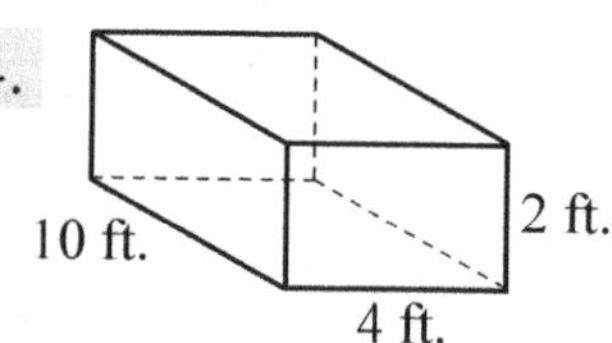

5.

7 cm.
13 cm.
3 cm.

6.

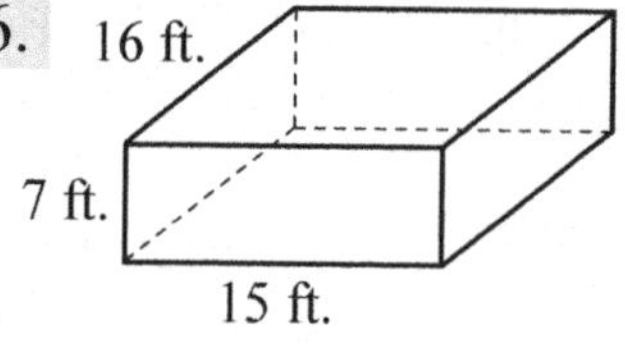

7.

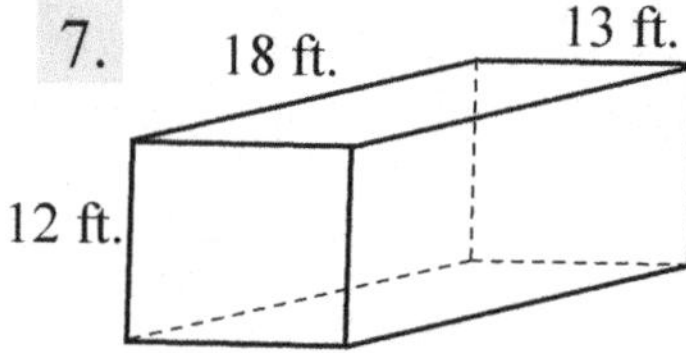

1.
2.
3.
4.
5.
6.
7.

Score

Problem Solving

A plane averages 450 miles per hour for the first four hours of a trip and 550 miles per hour for the next five hours. How many miles did the plane travel in all?

Review Exercises

1. Find the surface area.

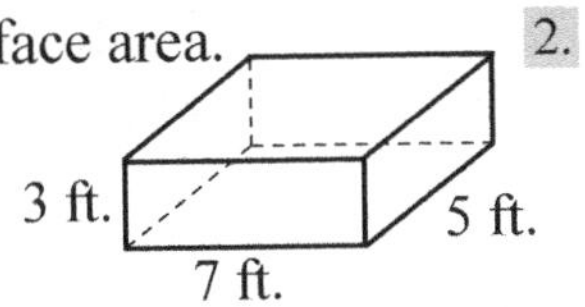

2. Find the volume.

4 ft. 8 ft. 6 ft.

3. What is the measure of ∠ BCD?

B 38° C D

4. What is the measure of ∠ DEF?

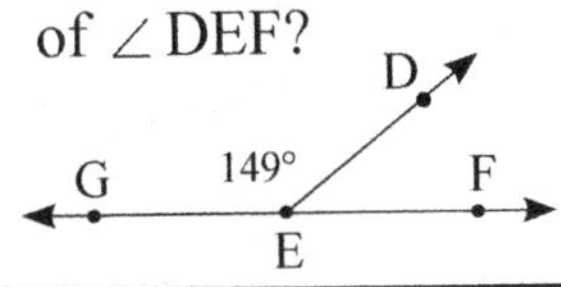

5. What is the measure of ∠ LMN?

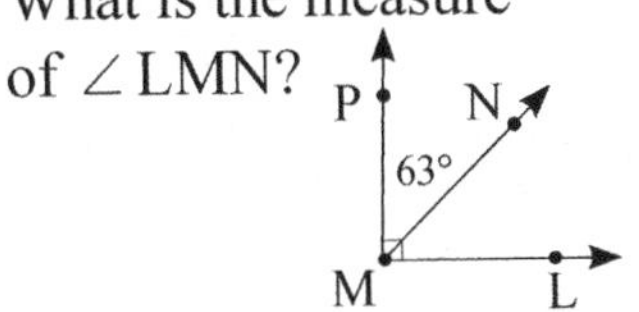

6. Find the circumference.

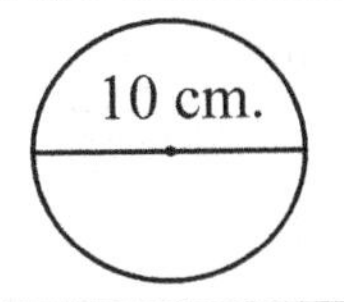

Helpful Hints

Use what you have learned to solve the following problems.
* Remember, volume is expressed in cubic units.
Examples: cm^3, ft^3, in^3

Find the volume for each figure. 1. Sketch a rectangular prism or cube. 2. Using the formula, substitute the values. 3. Find the volume.

S1. **Rectangular Prism**
l = 12 in.
w = 6 in.
h = 5 in.

S2. **Cube**
Each edge 7 ft.

1. **Rectangular Prism**
l = 7 ft.
w = 8 ft.
h = 10 ft.

2. **Cube**
Each edge 9 in.

3. **Rectangular Prism**
l = 24 ft.
w = 10 ft.
h = 5 ft.

4. **Cube**
Each edge 11 cm.

5. **Cube**
Each edge 16 in.

6. **Rectangular Prism**
l = 22 in.
w = 10 in.
h = 10 in.

7. **Rectangular Prism**
l = 7 ft.
w = 8 ft.
h = 5 ft.

1.
2.
3.
4.
5.
6.
7.
Score

Problem Solving

Mr. Ramos is purchasing vehicles for his company. He is purchasing three cars for $22,000 each and two trucks for $25,500 each. What will be the total cost of the vehicles?

Review Exercises

1. Find the surface area.

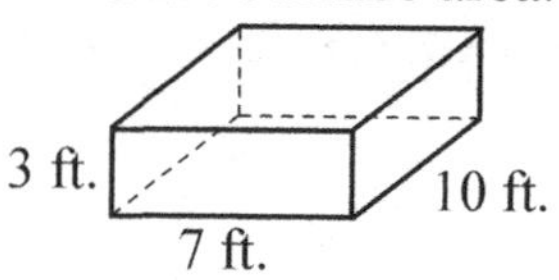

2. Find the volume.

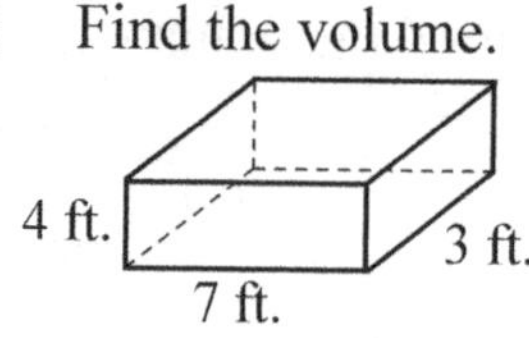

3. Find the surface area of the cube.

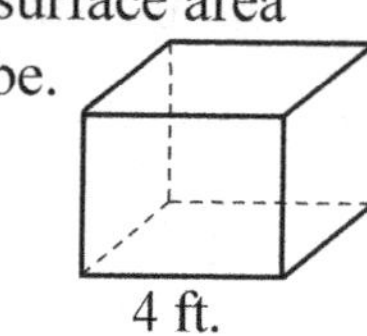

4. Find the volume of the cube.

4 ft.

5. Find the perimeter.

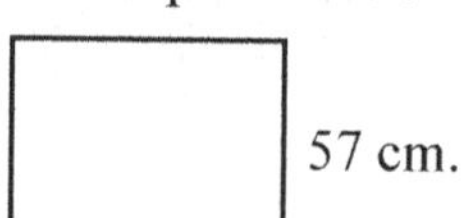

6. Find the area.

19 ft.

Helpful Hints

Use what you have learned to solve the following problems.

* Refer to previous "Helpful Hints" sections if necessary.
* Surface area is expressed in square units.
* Volume is expressed in cubic units.

Find the surface area or volume of each cube and rectangular prism.

S1. Find the volume.

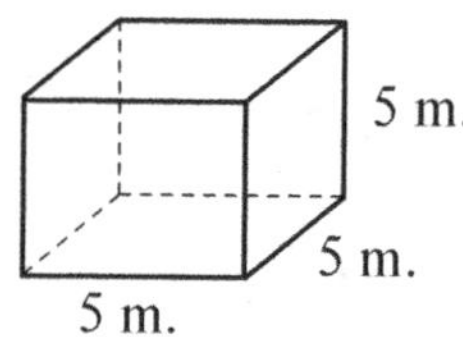

S2. Find the surface area.

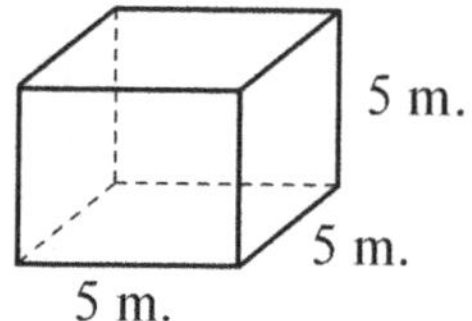

1. Find the surface area.

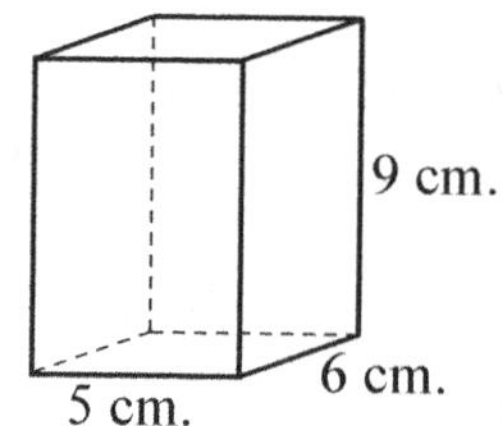

2. Find the volume.

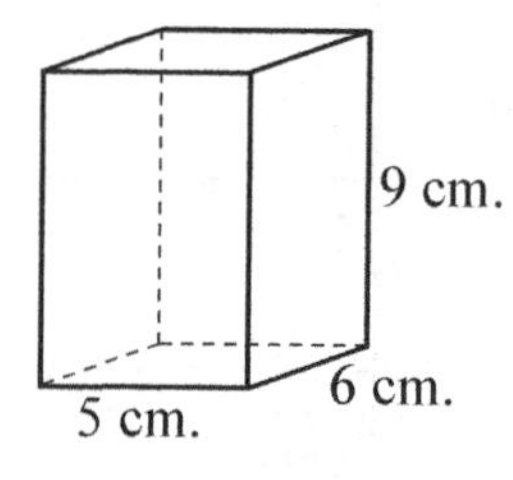

3. Find the volume.

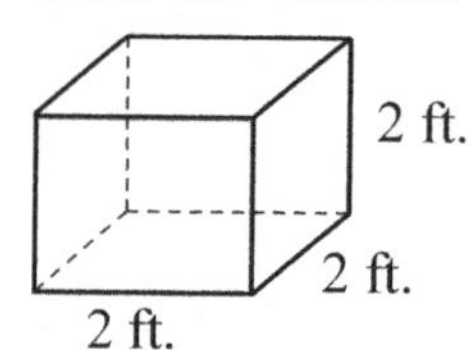

4. Find the surface area.

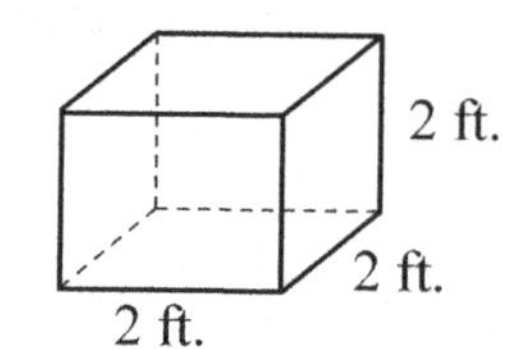

5. Find the surface area.

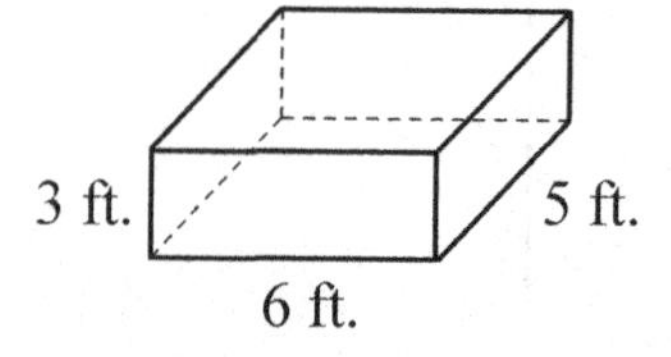

6. Find the volume.

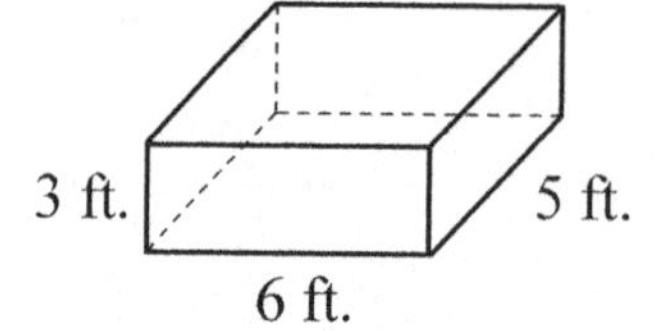

7. Find the volume.

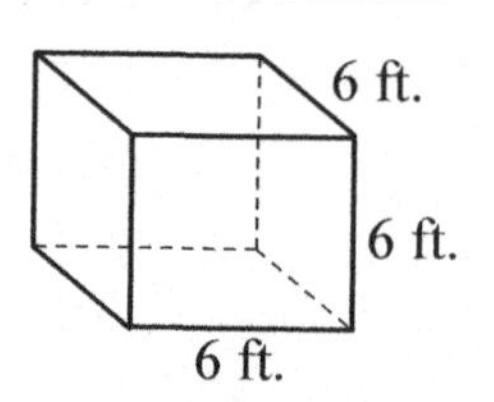

1.

2.

3.

4.

5.

6.

7.

Score

Problem Solving

Sonia wants to buy a cell phone for 75 dollars. She has 45 dollars and is saving another 5 dollars per day. How many days will it take her to save enough to buy the cell phone?

Review Exercises

For 1 - 6 find the area of each figure.

1. 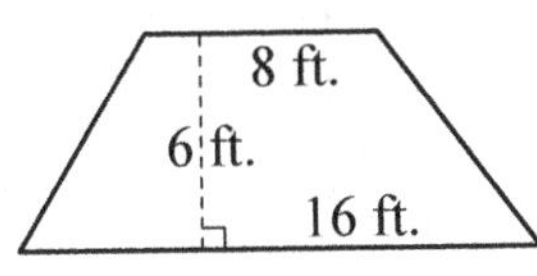

2. 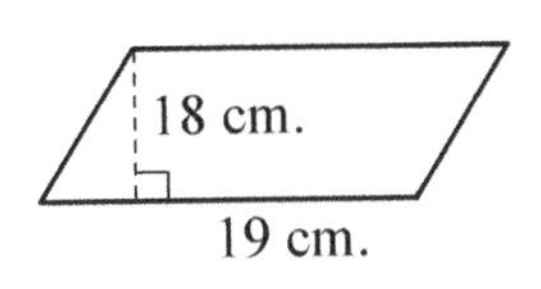

3. 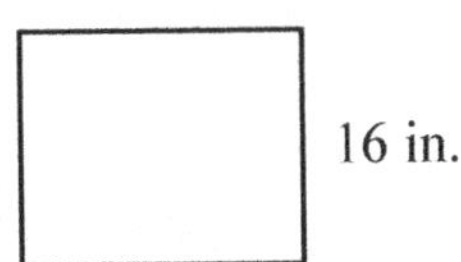

4. 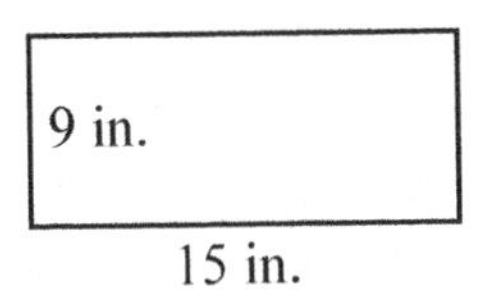

5. 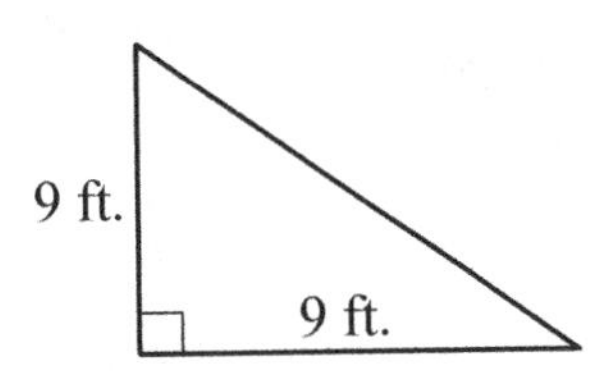

6. 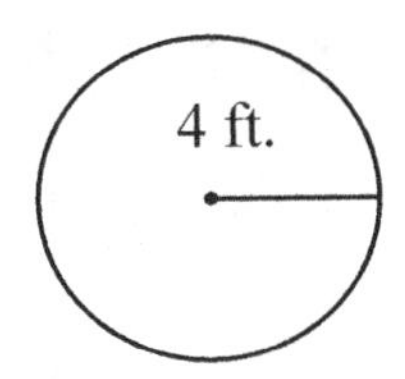

Helpful Hints

Use what you have learned to solve the following problems.

* Refer to previous "Helpful Hints" sections if necessary.
* Surface area is expressed in square units.
* Volume is expressed in cubic units.

Solve each of the following. Make a sketch. Next, substitute values. Finally, solve the problem.

S1. Find the volume of a rectangular prism with a length of 12 ft., width of 9 ft., and height of 7 ft.

S2. Find the surface area of a cube with edges of 4 inches in length.

1. A shed with a flat roof is 10 feet long, 8 feet wide, and 7 feet tall. What is the surface area of the shed?

2. A gift box in the shape of a cube has edges of 9 inches. How many square inches of wrapping paper is needed to cover the box?

3. An aquarium is 15 inches long, 10 inches wide, and 10 inches high. What is the volume of the aquarium?

4. Gill is going to paint the outside of a box which is six feet long, four feet wide, and five feet high. How many square feet of surface area will he paint?

5. The volume of a rectangular prism is 2,184 in.^{3}. If the length is 12 in. and the height is 14 in., what is the width?

1.

2.

3.

4.

5.

Score

Problem Solving

Nathan worked 1,840 hours last year. If his hourly pay was 18 dollars per hour, how much did he earn last year?

Final Review

Use the figure to answer problems 1 - 10.

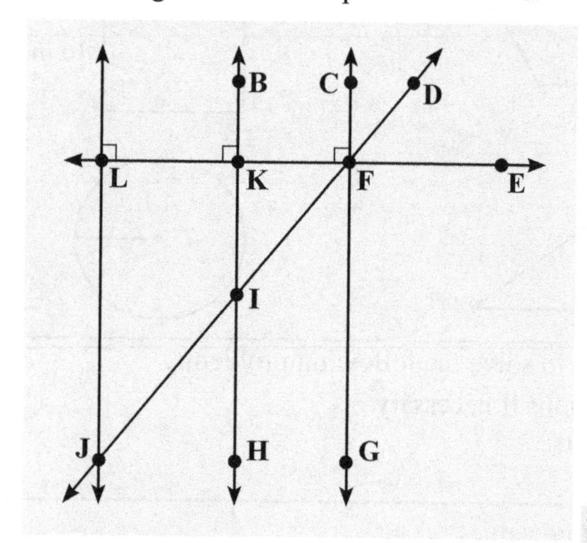

1. Name two parallel lines.
2. Name two perpendicular lines.
3. Name three line segments.
4. Name three rays.
5. Name two acute angles.
6. Name one triangle.
7. Name two obtuse angles.
8. Name one straight angle.
9. Name two right angles.
10. Name one trapezoid.

Use the figures below to answer problems 11 - 12.

Triangle A

Triangle B

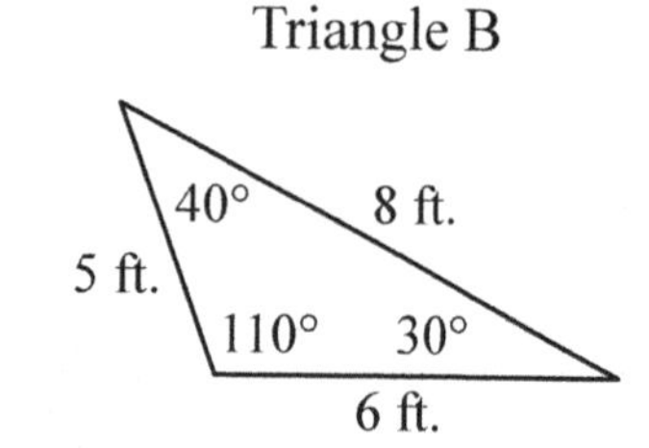

11. Classify triangle A by its sides and angles.
12. Classify triangle B by its sides and angles.

For problems 13 - 17 use the circle below.

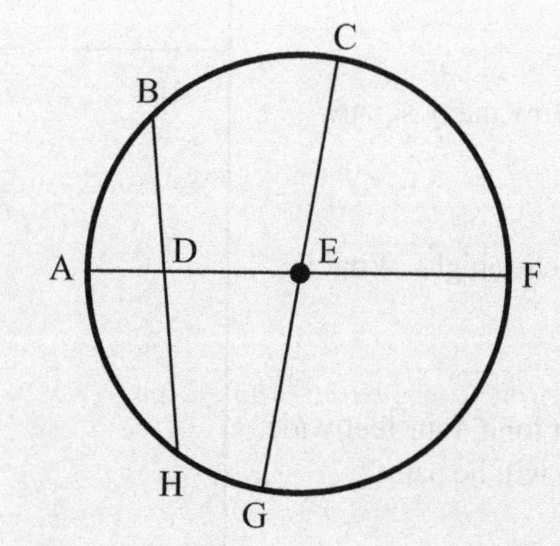

13. Name two diameters.
14. Name three radii.
15. Name two chords.
16. If $\overline{CG}$ is 24 ft., what is the length of $\overline{CE}$?
17. What part of the circle is $\overline{BH}$?

For problems 18 - 20 identify the shape and numbers of each part.

18.

Name:________
Faces:________
Edges:________
Vertices:________

19.

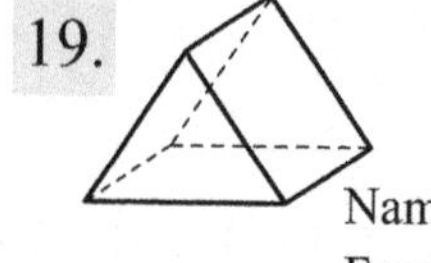

Name:________
Faces:________
Edges:________
Vertices:________

20.

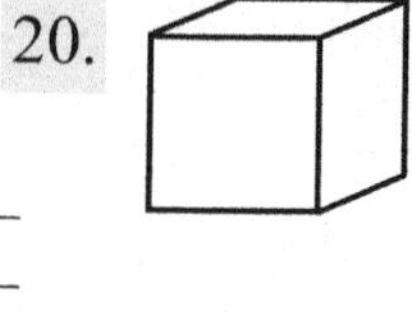

Name:________
Faces:________
Edges:________
Vertices:________

1.
2.
3.
4.
5.
6.
7.
8.
9.
10.
11.
12.
13.
14.
15.
16.
17.
18.
19.
20.

Final Test Concepts

Use the figure to answer problems 1 - 10.

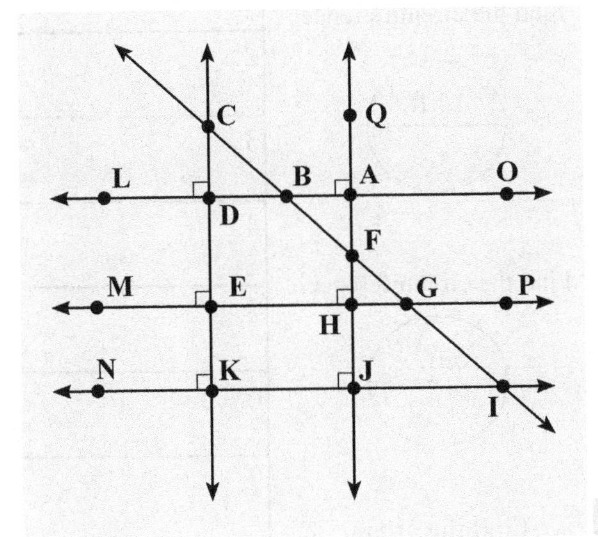

1. Name two parallel lines.
2. Name two perpendicular lines.
3. Name an obtuse, an acute and a right angle.
4. Name two rays.
5. Name two triangles.
6. Name a trapezoid.
7. Name two line segments.
8. Name one pentagon.
9. Name one straight angle.
10. Name two congruent line segments.

Use the figures below to answer problems 11 - 13.

Triangle A

5 ft. 85° 7 ft.

60° 35°

9 ft.

Triangle B

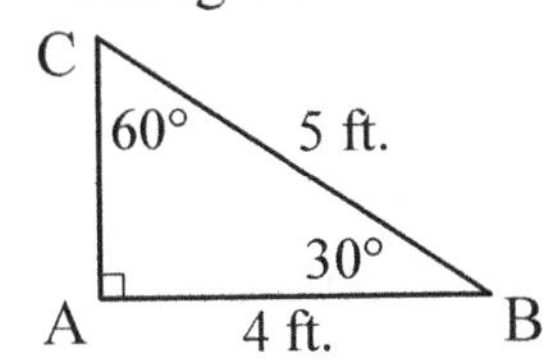

11. Classify triangle A by its sides and angles.
12. Classify triangle B by its sides and angles.
13. In triangle B, what part of the triangle is $\overline{CB}$?

14. What polygon has one pair of parallel sides?

15. Name three polygons that have two pairs of parallel sides.

16. What polygon has eight sides?

17. What term is given to a polygon with all its angles congruent and all its sides congruent?

For problems 18 - 20 identify the shape and numbers of each part.

18. 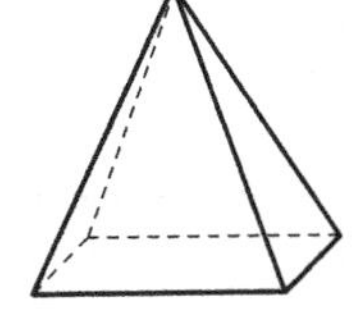

Name:________
Faces:________
Edges:________
Vertices:______

19. 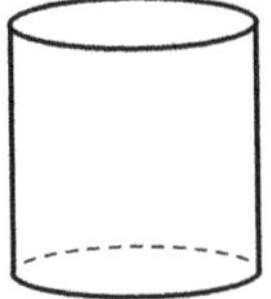

Name:________
Faces:________
Edges:________
Vertices:______

20. 

Name:________
Faces:________
Edges:________
Vertices:______

1. ________
2. ________
3. ________
4. ________
5. ________
6. ________
7. ________
8. ________
9. ________
10. ________
11. ________
12. ________
13. ________
14. ________
15. ________
16. ________
17. ________
18. ________
19. ________
20. ________

Final Review

1. Find the perimeter.

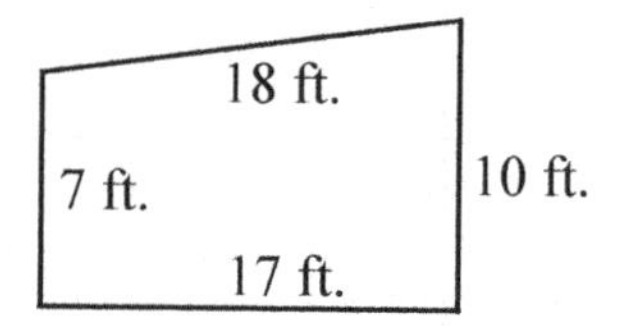

2. Find the perimeter.

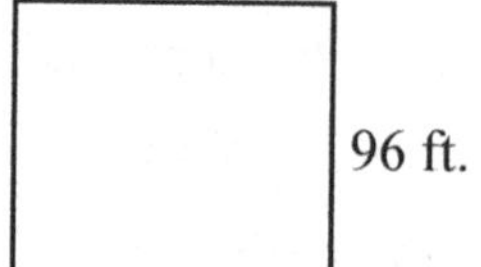

3. Find the circumference.

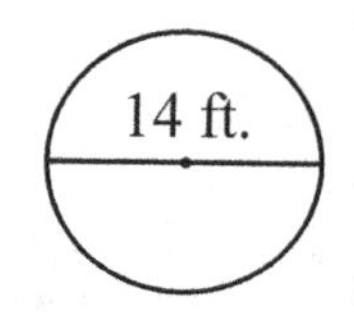

4. Find the perimeter of a regular hexagon with sides 29 ft.

5. Find the perimeter.

18 ft.
39 ft.

6. Find the circumference.

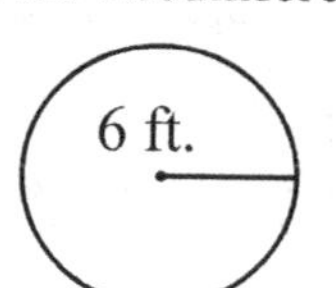

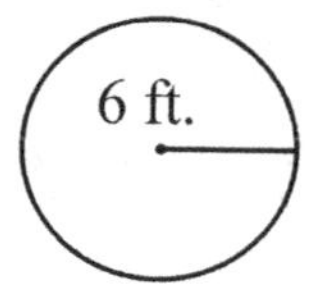

7. Find the area.

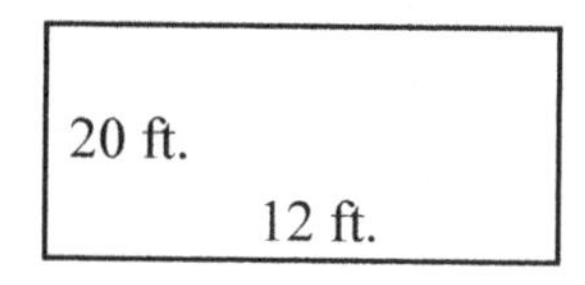

8. Find the area.

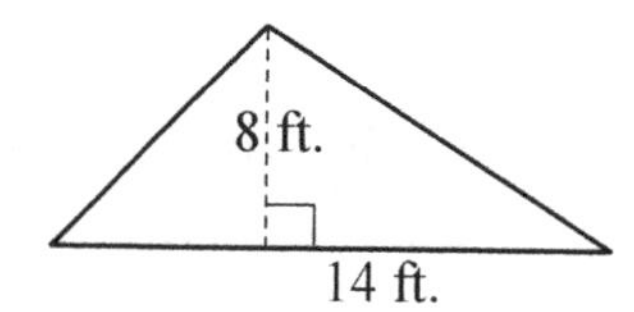

9. Find the area.

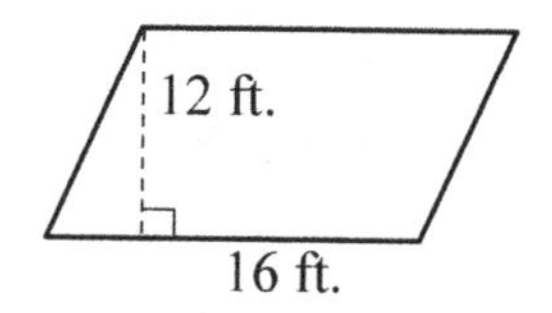

10. Find the area.

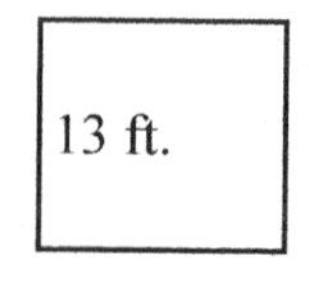

11. Find the area.

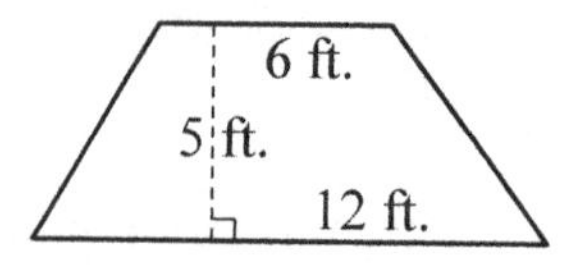

12. Find the area.

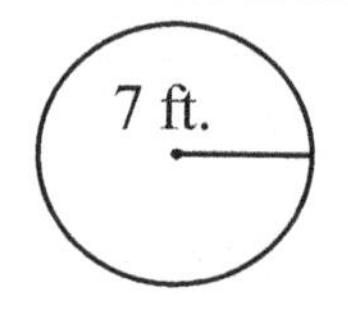

13. Find the area.

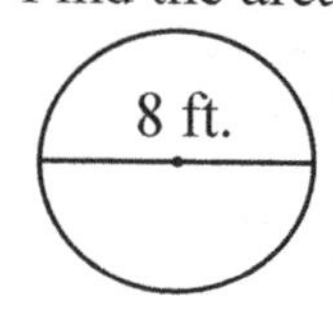

14. Find the area.

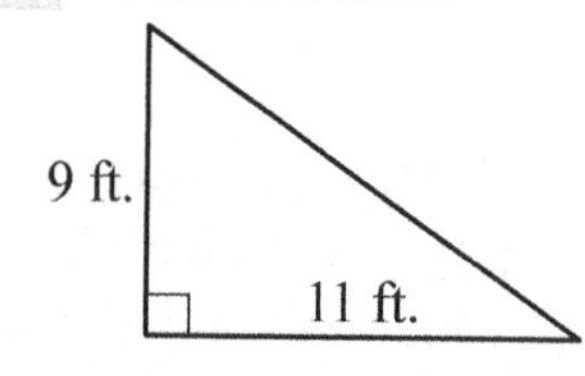

15. Find the area.

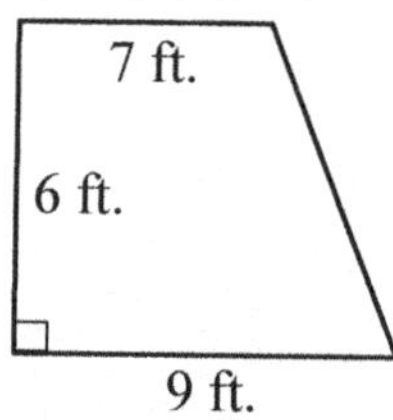

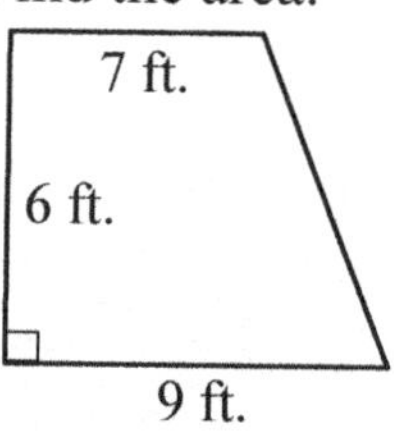

16. Find the area of a square with sides 17 ft.

17. Find the area of a rectangle with length 18 cm. and width 12 cm.

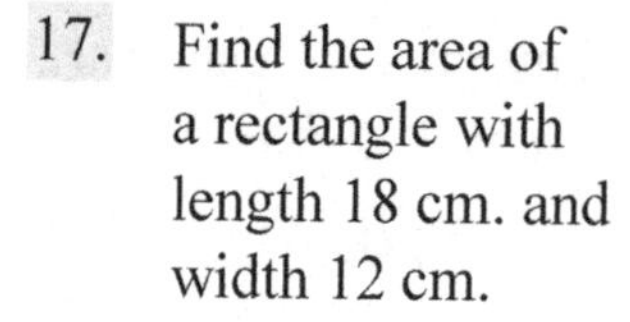

18. Find the perimeter of a regular octagon with sides 19 ft.

19. Find the surface area.

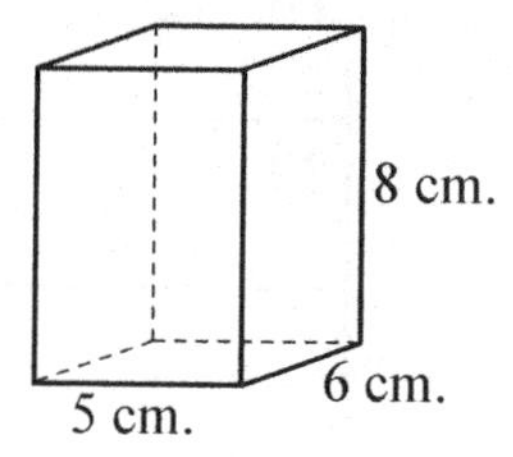

20. Find the volume.

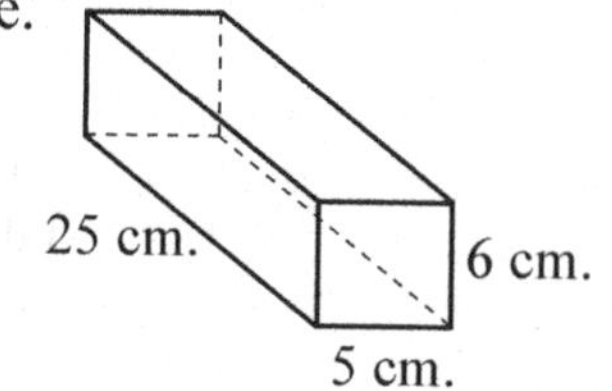

1. ______
2. ______
3. ______
4. ______
5. ______
6. ______
7. ______
8. ______
9. ______
10. ______
11. ______
12. ______
13. ______
14. ______
15. ______
16. ______
17. ______
18. ______
19. ______
20. ______

Final Test

1. Find the perimeter.

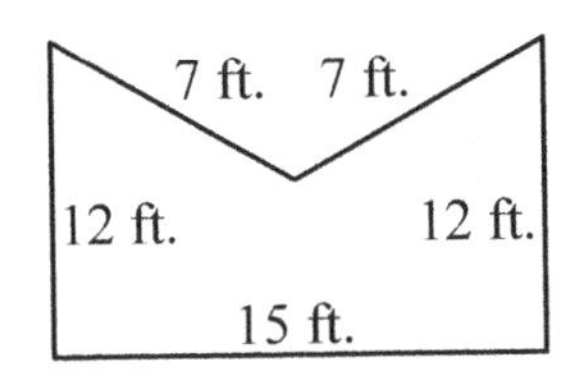

2. Find the circumference.

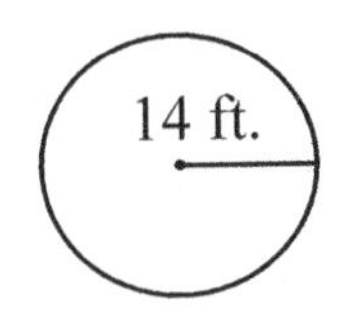

3. Find the perimeter.

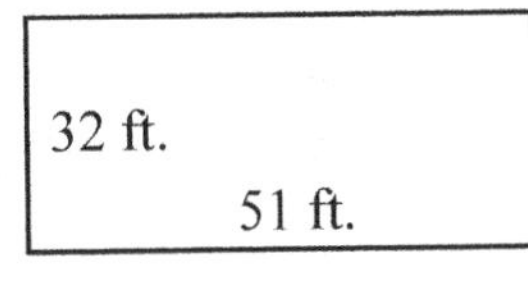

4. Find the circumference.

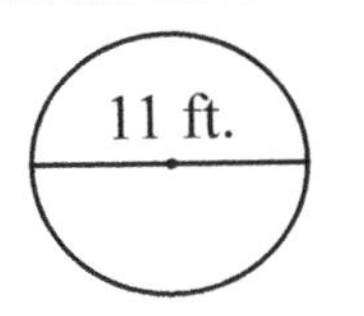

5. Find the perimeter.

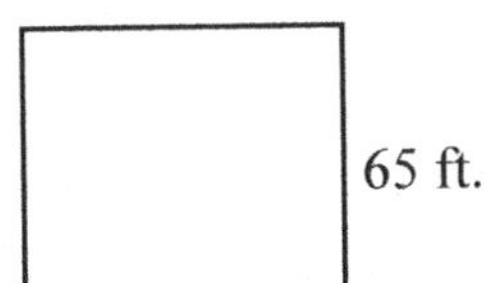

6. Find the perimeter of a regular decagon with sides 29 ft.

7. Find the area.

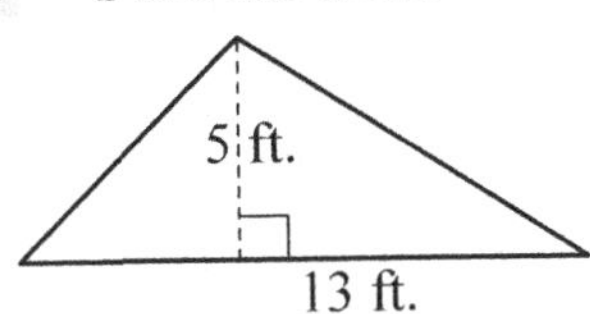

8. Find the area.

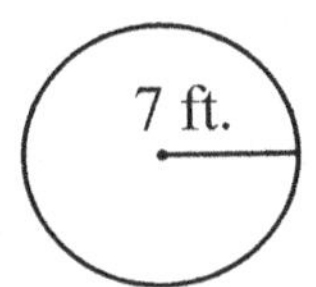

9. Find the area.

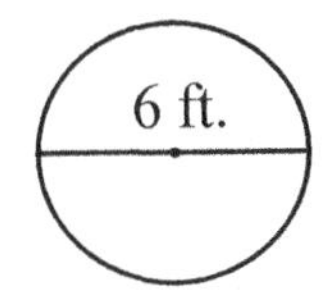

10. Find the area.

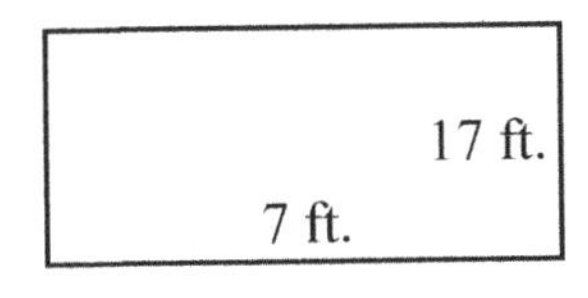

11. Find the area.

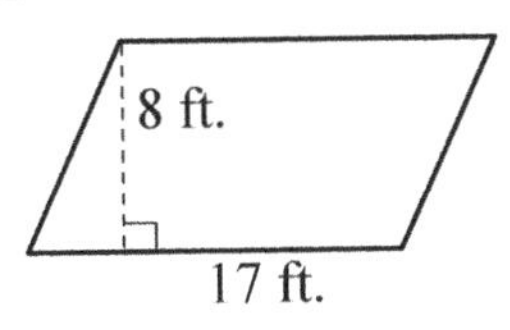

12. Find the area.

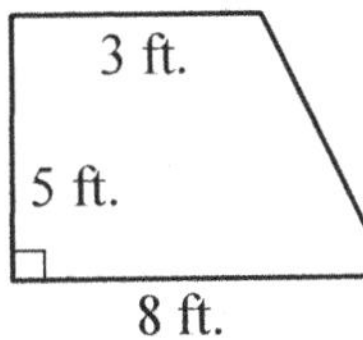

13. Find the area of a trapezoid.
    B = 12 ft., b = 6 ft., h = 3ft.

14. Find the area of a square with sides 16 ft.

15. Find the area of a triangle with b = 7 ft. and h = 5 ft.

16. Find the area.

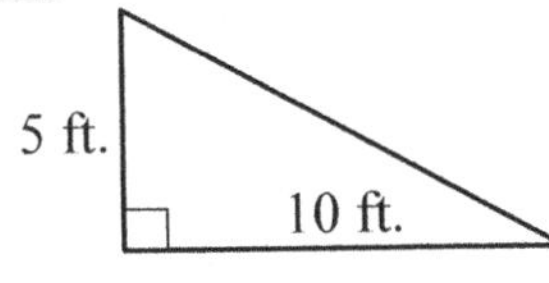

17. Find the area.

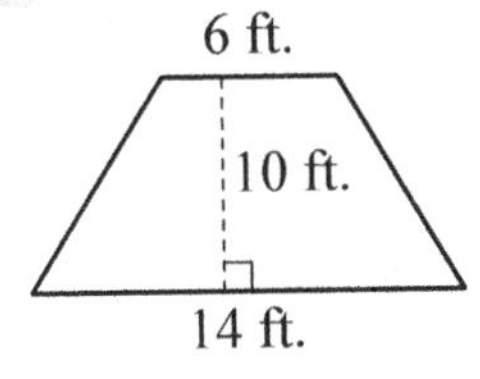

18. Find the area of a rectangle with length 60 ft. and width 12 ft.

19. Find the surface area.

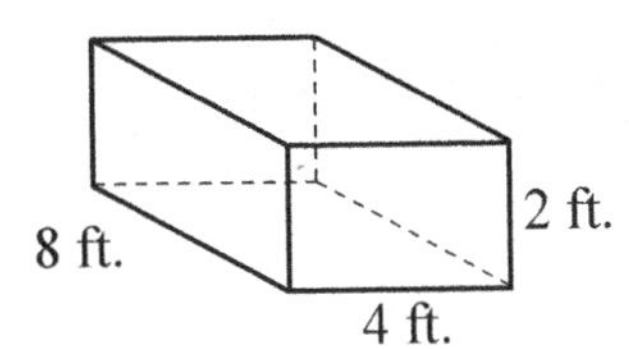

20. Find the volume.

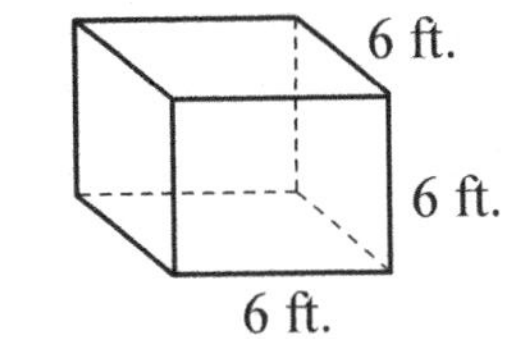

1.
2.
3.
4.
5.
6.
7.
8.
9.
10.
11.
12.
13.
14.
15.
16.
17.
18.
19.
20.

Section 2

Problem Solving

Review Exercises

Note to students and teachers: This section will include necessary review problems from all topics covered in this book. Here are some simple problems with which to get started.

1. $\begin{array}{r} 336 \\ 27 \\ +\ 242 \\ \hline \end{array}$

2. $\begin{array}{r} 752 \\ -\ 68 \\ \hline \end{array}$

3. 725 + 242 + 163 =

4. $\begin{array}{r} 324 \\ \times 6 \\ \hline \end{array}$

5. $\begin{array}{r} 5{,}003 \\ \times 6 \\ \hline \end{array}$

6. 7 x 6,382 =

Helpful Hints

Bar graphs are used to compare information.

1. Read the title.
2. Understand the meaning of the numbers. Estimate, if necessary.
3. Study the data.
4. Answer the questions, showing work if necessary.

Use the information in the graph to answer the questions.

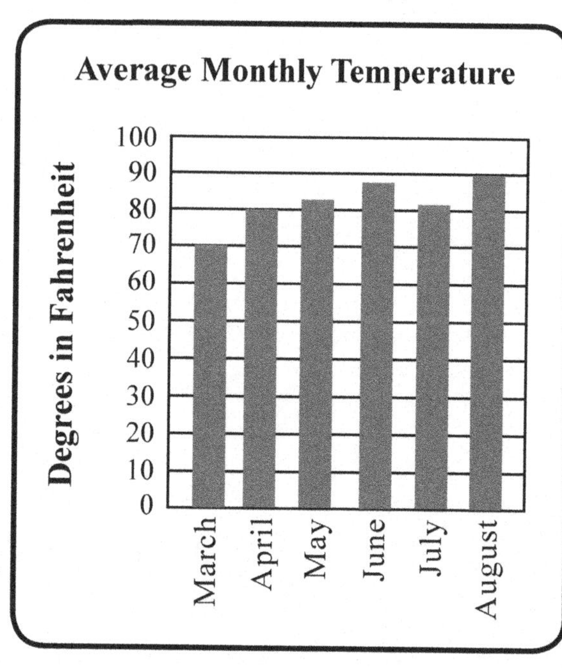

S1. Which month had the second-lowest average temperature?

S2. How many degrees cooler was the average temperature in April than in August?

1. In which month was the average temperature 81°?

2. In which month did the average temperature drop from the previous month?

3. Which month had the second-highest average temperature?

4. How much warmer was the average temperature in August than in May?

5. For what month did the average temperature rise the most from the previous month?

6. Which months had average temperatures less than July's average temperature?

7. Which two month's average temperatures were the closest?

8. The coolest day in August was 77°. How much less than the average temperature was this?

9. What was the increase in average temperature from May to June?

10. Which months had an average temperature less than May?

1.
2.
3.
4.
5.
6.
7.
8.
9.
10.
Score

Review Exercises

1. 555
   77
   + 888

2. 703
   − 276

3. 642
   x 7

4. 42
   x 25

5. 47
   x 20

6. 76
   x 33

Helpful Hints

1. Read the title.
2. Understand the meaning of the numbers. Estimate, if necessary.
3. Study the data.
4. Answer the questions, showing work if necessary.

Use the information in the graph to answer the questions.

Population of Cities in Riverdale County

Mayfield
Springdale
Auberry
Lincoln
Sun City
Winston

0 1 2 3 4 5 6 7 8

Number of People in 100's

S1. Which two cities had the same population?

S2. What is the combined population of Mayfield and Lincoln?

1. Within 3 years, the population of Lincoln is expected to double. What will its population be in 3 years?

2. How many more people live in Springdale than in Auberry?

3. What is the difference in population between the largest and smallest cities?

4. What is the total population of Riverdale County?

5. How many more people live in Sun City than in Mayfield?

6. To reach a population of 900, by how much must Mayfield grow?

7. What is the total population of the two largest cities?

8. How many people must move to Winston before its population is equal to that of Springdale?

9. Which city has nearly double the population of Auberry?

10. What is the total population of all cities whose population is less than 500?

1.
2.
3.
4.
5.
6.
7.
8.
9.
10.
Score

Review Exercises

1. 3,126 x 5

2. 64 x 23

3. 203 x 47

4. 164 x 23

5. 3 ⟌ 617

6. 5 ⟌ 236

Helpful Hints

Line graphs are used to show changes and relationships between quantities.

1. Read the title.
2. Understand the meaning of the numbers. Estimate if necessary.
3. Study the data.
4. Answer the questions, showing work if necessary.

Use the information in the graph to answer the questions.

S1. What was John's score on Test 5?

S2. How much better was John's score on Test 7 than on Test 3?

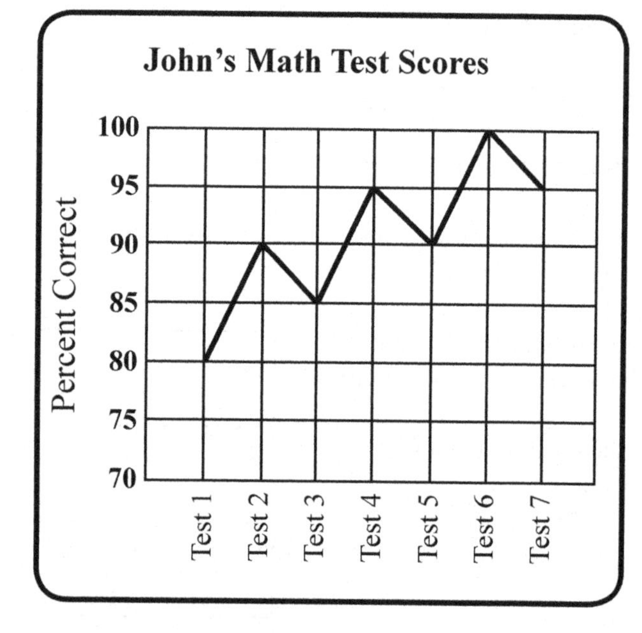

1. On which three tests did John score the highest?

2. What is the difference between his highest and lowest score?

3. Find John's average score by adding all his scores and dividing by the number of scores.

4. How many scores were below John's average score?

5. What are his two lowest scores?

6. What is the average of his highest and lowest scores?

7. How much higher was Test 4 than Test 5?

8. What was the difference between his highest score and his second lowest score?

9. How many test scores were improvements over the previous test?

10. Did John's progress generally improve or get worse?

1.
2.
3.
4.
5.
6.
7.
8.
9.
10.
Score

Review Exercises

1. $\begin{array}{r} 3{,}763 \\ 472 \\ +\ 5{,}637 \\ \hline \end{array}$

2. $\begin{array}{r} 5{,}016 \\ -\ 738 \\ \hline \end{array}$

3. $\begin{array}{r} 435 \\ \times\ 26 \\ \hline \end{array}$

4. $7\overline{)1{,}407}$

5. $6\overline{)2{,}397}$

6. $8\overline{)1{,}445}$

Helpful Hints

1. Read the title.
2. Understand the meaning of the numbers. Estimate, if necessary.
3. Study the data.
4. Answer the questions, showing work if necessary.

Use the information in the graph to answer the questions.

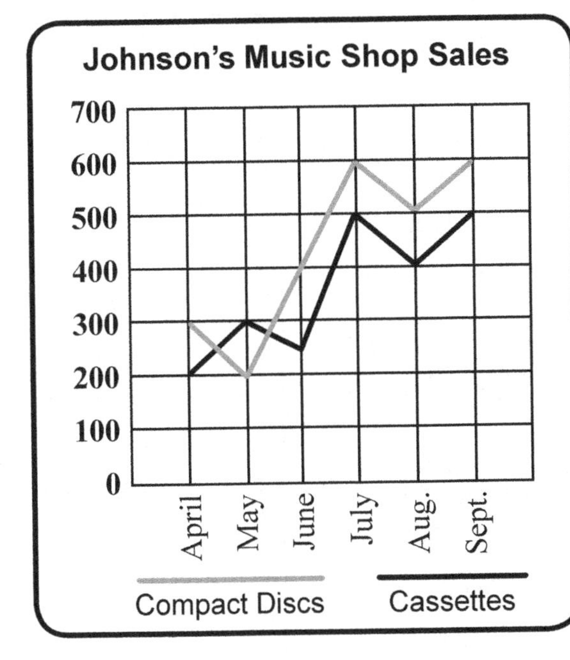

S1. In what months were the most cassettes sold?

S2. How many more compact discs than cassettes were sold in August?

1. What was the total number of compact discs sold in May and August?

2. How many more compact discs were sold in July than in August?

3. In which month were more cassettes sold than compact discs?

4. During which month did compact discs outsell cassettes by the most?

5. In September, how many more compact discs were sold than cassettes?

6. Which two months had the highest total sales?

7. What was the total number of compact discs and cassettes sold in September?

8. What was the increase in sales of compact discs between May and June?

9. What was the decrease in sales of compact discs between July and August?

10. What were the two lowest total sales months?

1.
2.
3.
4.
5.
6.
7.
8.
9.
10.
Score

Review Exercises

1. 664×23

2. $5{,}000 - 856$

3. $435 + 75 + 61 + 42 =$

4. $30 \overline{)96}$

5. $50 \overline{)765}$

6. $9 \overline{)1{,}809}$

Helpful Hints	A circle graph shows the relationship between the parts to the whole and to each other.	1. Read the title. 2. Understand the meaning of the numbers. Estimate if necessary. 3. Study the data. 4. Answer the questions, showing work if necessary.

Use the information in the graph to answer the questions.

S1. What percent of the family budget is spent for food?

S2. After the car payment and house payment are paid, what percent of the budget is left?

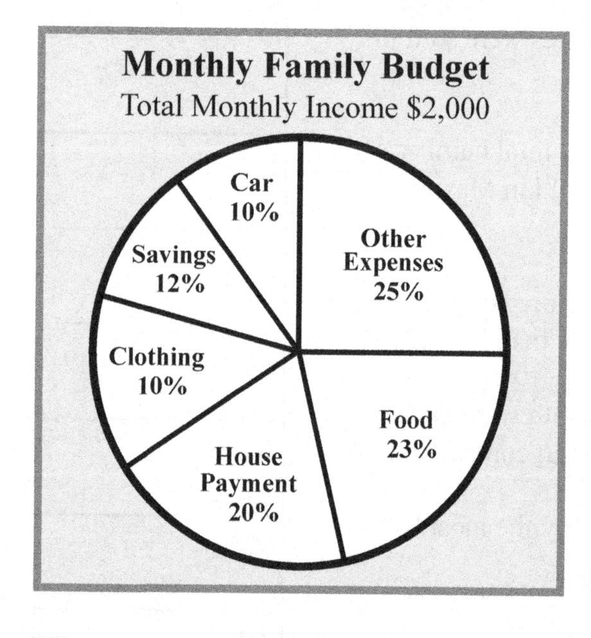

1. What percent of the budget is used to pay for food and clothing?

2. How much is spent on food each month? (Hint: Find 23% of $2,000.)

3. How many dollars are spent on clothing?

4. How many dollars was the house payment?

5. What percent of the budget did the three largest items represent?

6. What percent of the budget is for savings, clothing, and the car payment?

7. In twelve months, what is the total income?

8. What percent of the budget is left after the house payment has been made?

9. Which two items require the same part of the budget?

10. Which part of the budget would pay for medical expenses?

1.
2.
3.
4.
5.
6.
7.
8.
9.
10.
Score

Review Exercises

1. $2\overline{)76}$

2. $725 \div 5 =$

3. $80\overline{)905}$

4. $70\overline{)697}$

5. $70\overline{)8,196}$

5. $30\overline{)6,152}$

Helpful Hints

Circle graphs can be used to show fractional parts.

1. Read the title.
2. Understand the meaning of the numbers.
3. Study the data.
4. Answer the questions.

Use the information in the graph to answer the questions.

S1. What fraction of the day does Jane play?

S2. What fraction of Jane's day is used for school?

Jane's School Day

1. How many more hours does Jane sleep per day than play?

2. How many hours of homework does Jane have in a week (Monday through Friday)?

3. How many hours per day are school-related activities?

4. What fraction of the day is spent for school, homework, and chores?

5. What fraction of the day does Jane spend for school, sleep, and chores?

6. How many hours does Jane spend in school in 3 weeks?

7. If Jane goes to bed at 9:00 p.m., what time does she get up in the morning?

8. If school starts at 8:30 a.m., what time is school dismissed?

9. How many hours per week does Jane spend in school and on homework?

10. What fractional part of the day does Jane play and have meals?

1.
2.
3.
4.
5.
6.
7.
8.
9.
10.
Score

Review Exercises

1. 364 + 27 + 256 + 427 =

2. 7,000 − 1,356 =

3. 365 x 246 =

4. 22 ⟌463

5. 31 ⟌651

6. 18 ⟌379

Helpful Hints	Picture graphs are another way to compare statistics.	1. Read the title. 2. Understand the meaning of the numbers. Estimate, if necessary. 3. Study the data. 4. Answer the questions.

Use the information in the graph to answer the questions.

Bikes Made By Street Bike Company

Year	Bikes
1986	🚲 🚲 🚲
1987	🚲 🚲 🚲 🚲
1988	🚲 🚲 🚲 ½🚲
1989	🚲 🚲 🚲 🚲 🚲 🚲
1990	🚲 🚲 🚲 🚲 🚲 ½🚲
1991	🚲 🚲 🚲 🚲 🚲 🚲 🚲 🚲

Each 🚲 represents 1,000 bikes.

S1. How many bikes were made in 1989?

S2. How many more bikes were made in 1991 than in 1988?

1. Which year produced twice as many bikes as 1986?

2. What was the total number of bikes that the company produced in 1990 and 1991?

3. Which two years did the company make the most bikes?

4. 1992 is reported to be double the production of 1988. How many bikes are to be produced in 1992?

5. What is the total number of biked produced in 1986 and 1991?

6. It cost $50 to make a bike in 1986. How much did the company spend making bikes that year?

7. The cost to make bikes jumped to $75 in 1991. How much did the company spend making bikes in 1991?

8. What is the difference in bikes produced in 1986 and 1991?

9. What is the total number of bikes made during the company's 3 most productive years?

10. One half of the bikes made in 1989 were ladies' style. How many ladies' bikes were made in 1989?

Answers
1.
2.
3.
4.
5.
6.
7.
8.
9.
10.
Score

Review Exercises

1. $3\overline{)52}$
2. $7\overline{)1{,}693}$
3. $30\overline{)697}$
4. $60\overline{)4{,}287}$
5. $22\overline{)7{,}695}$
6. $38\overline{)8{,}561}$

Helpful Hints

1. Read the title.
2. Understand the meaning of the numbers. Estimate if necessary.
3. Study the data.
4. Answer the questions, showing work if necessary.

Use the information in the graph to answer the questions.

S1. Which work week was the longest?

S2. How many hours shorter was the work week in 1960 than in 1990?

America's Work Week

1950

1960

1970

1980

1990

Each symbol represents 10 hours

1. How many hours long was the work week in 1970?
2. How many hours did the work week increase between 1980 and 1990?
3. If the average employee works 50 weeks per year how many hours did he work in 1950?
4. Which year's work week was approximately 38 hours?
5. Which years had the 2 shortest work weeks?
6. How many hours less was the work week in 1950 than 1980?
7. If the work week is 5 days, what was the average number of hours worked per day in 1950?
8. In 1990, if any employee decided to work 4 days per week, what would his average number of hours be per day?
9. Any work time over 40 hours is overtime. What was the average worker's weekly overtime in 1970?
10. What is the difference between the longest and the shortest work week?

1.
2.
3.
4.
5.
6.
7.
8.
9.
10.
Score

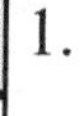

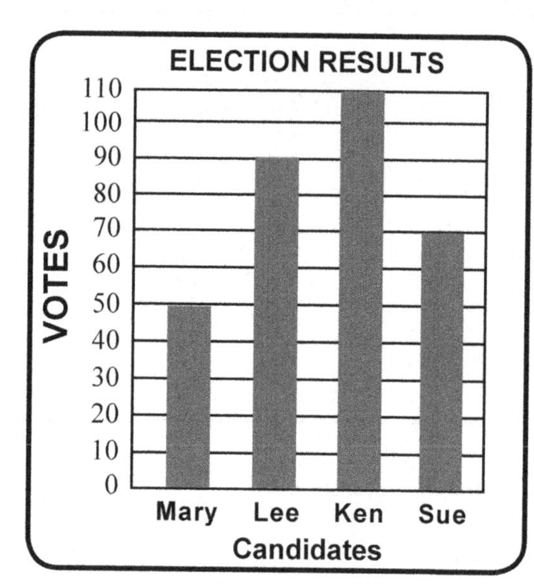

1. Which candidate got the most votes?
2. How many votes did Sue receive?
3. How many more votes did Lee get than Mary?
4. Together, how many votes did Lee and Ken receive?
5. Who got the third most votes?

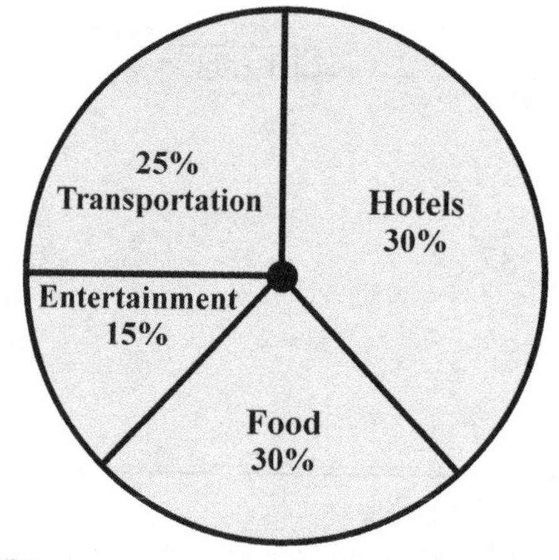

6. What percent is spent on hotels?
7. What percent of the budget is spent on transportation?
8. What percent is spent altogether on transportation and hotels?
9. What is the second largest part of the budget?
10. What percent of the budget is used for entertainment?

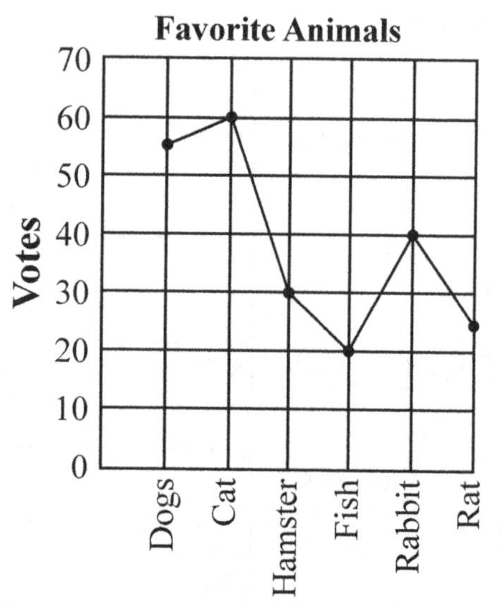

11. How many picked cats as their favorite animal?
12. What was the favorite animal?
13. What was the least favorite animal?
14. How many more people picked cats than hamsters?
15. What was the second favorite animal?

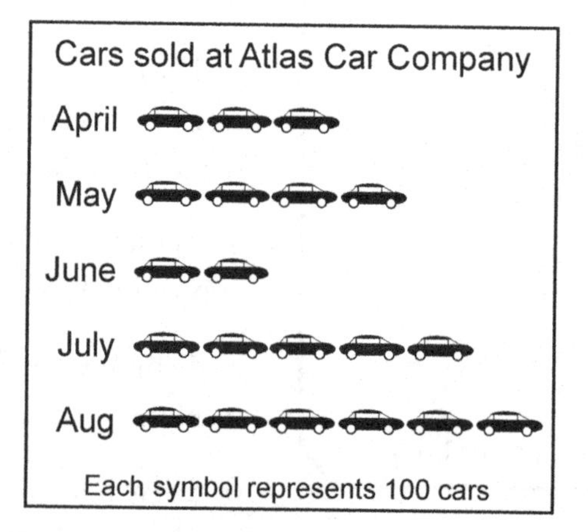

16. How many cars were sold in May?
17. Which month had the highest car sales?
18. How many more cars were sold in August than in June?
19. What was the total number of cars sold in April and June?
20. Which two months had the highest sales?

1.
2.
3.
4.
5.
6.
7.
8.
9.
10.
11.
12.
13.
14.
15.
16.
17.
18.
19.
20.

Score

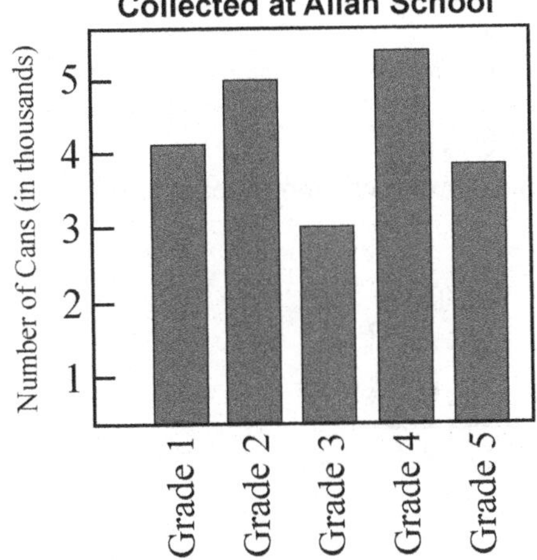

1. Which grade collected the most cans?
2. How many more cans did grade 4 collect than grade 1?
3. How many cans did grades 4 and 5 collect altogether?
4. How many cans were collected in all?
5. Which grade collected the second most number of cans?

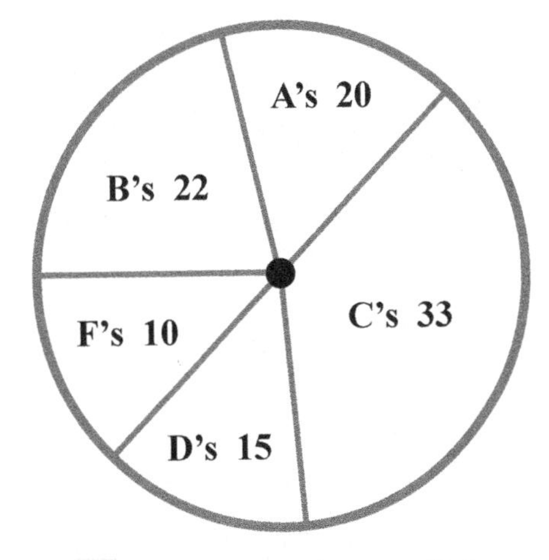

6. How many students got A's?
7. How many more C's than B's were there?
8. How many science students were there in all?
9. How many more C's than B's were there than A's and B's?
10. Which was the second largest group of grades?

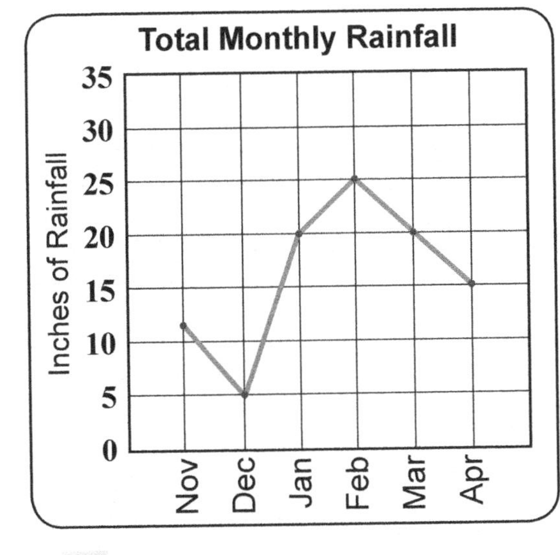

11. How many inches of rain fell in February?
12. How many more inches of rain fell in January than in November?
13. Which month's rainfall increased the most from the previous month?
14. What was the total amount of rain for February, March, and April?
15. Which month's rainfall decreased the most from the previous month?

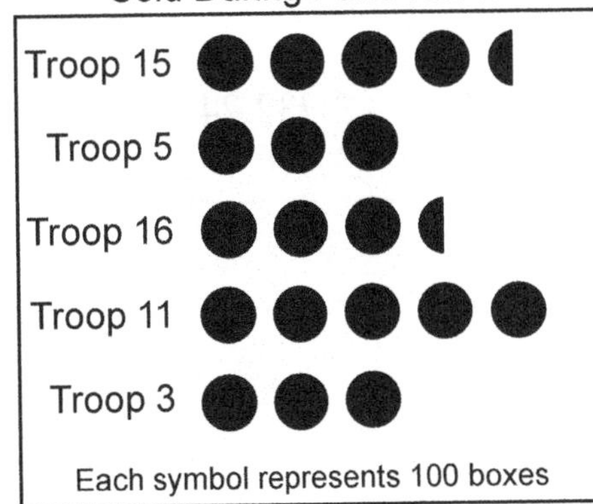

16. How many boxes of cookies did Troop 11 sell?
17. Which troop sold the second most cookies?
18. How many more boxes did Troop 11 sell than Troop 5?
19. If Troop 3 sells three times as many boxes of cookies next year than they did this year, how much will they sell next year?
20. How many boxes of cookies did Troop 5 and Troop 15 sell altogether?

1.
2.
3.
4.
5.
6.
7.
8.
9.
10.
11.
12.
13.
14.
15.
16.
17.
18.
19.
20.
Score

Addition:
1. Line up on the right.
2. Place commas when necessary.
3. Add the ones first.
4. Regroup when necessary.
5. "Sum" means to add.

Example:

$$\begin{array}{r} {}^{2\ 2\ 1} \\ 7{,}654 \\ 2{,}765 \\ 793 \\ +\ \ 64 \\ \hline 11{,}276 \end{array}$$

Subtraction:
1. Line up on the right.
2. Place commas when necessary.
3. Subtract the ones first.
4. Regroup when necessary.
5. "Difference" and "how much more" means to subtract.

Example:

$$\begin{array}{r} {}^{6\ 11\ 9} \\ 7{,}203 \\ -\ 3{,}654 \\ \hline 3{,}549 \end{array}$$

S1.
$$\begin{array}{r} 743 \\ 7{,}614 \\ 16{,}321 \\ +\ 5{,}032 \\ \hline \end{array}$$

S2.
$$\begin{array}{r} 5{,}001 \\ -\ 1{,}346 \\ \hline \end{array}$$

1.
$$\begin{array}{r} 346 \\ 25 \\ +\ 176 \\ \hline \end{array}$$

2.
$$\begin{array}{r} 716 \\ -\ 143 \\ \hline \end{array}$$

3.
$$\begin{array}{r} 4{,}216 \\ 764 \\ +\ 5{,}123 \\ \hline \end{array}$$

4.
$$\begin{array}{r} 3{,}732{,}246 \\ +\ 3{,}510{,}762 \\ \hline \end{array}$$

5.
$$\begin{array}{r} 7{,}101 \\ -\ 1{,}436 \\ \hline \end{array}$$

6. Find the difference between 1,964 and 768.

7. 17,023 – 13,605 =

8. Find the sum of 236, 742, and 867.

9. How much more is 763 than 147?

10. 72,163 + 16,432 + 1,963 =

1.
2.
3.
4.
5.
6.
7.
8.
9.
10.
Score

Multiplication:

1. Line up on the right.
2. Multiply ones first.
3. Multiply tens next.
4. Multiply hundreds last.
5. Add the products.
6. Place commas when necessary.

Example:

```
     243
   x 346
   -----
   1,458
   9,720
   72900
  ------
  84,078
```

Division:

1. Divide.
2. Multiply.
3. Subtract.
4. Divide again.
5. Remainders must be less than the divisor.

Example:

```
      216 r20
 32 | 6,932
     -64↓↓
      53
     -32
      212
     -192
       20
```

S1. 675 x 43

S2. 32 ⟌ 7,049

1. 76 x 3

2. 7,653 x 4

3. 627 x 36

4. 673 x 346

5. 3 ⟌ 425

6. 6 ⟌ 1697

7. 30 ⟌ 769

8. 40 ⟌ 3762

9. 42 ⟌ 8992

10. 28 ⟌ 1577

1.
2.
3.
4.
5.
6.
7.
8.
9.
10.
Score

Review Exercises

1. $\begin{array}{r} 246 \\ 78 \\ +\ 912 \\ \hline \end{array}$

2. $\begin{array}{r} 7{,}562 \\ -\ \ 399 \\ \hline \end{array}$

3. $\begin{array}{r} 709 \\ \times\ \ 3 \\ \hline \end{array}$

4. $\begin{array}{r} 17 \\ \times\ 23 \\ \hline \end{array}$

5. $2\overline{)117}$

6. $9\overline{)1{,}763}$

Helpful Hints

1. Read the problem carefully.
2. Find the important facts and numbers.
3. Decide what operation to use.
4. Solve the problem and label it with a word or short phrase.

S1. There are four classes with enrollments of 32, 35, 33, and 38. What is the total enrollment?

S2. A family taking a vacation traveled 316 miles per day for five days. How many miles did they travel altogether?

1. The attendance at the concert last year was 7,652. This year 8,043 attended. What was the increase in attendance?

2. Six friends earned a total of 834 dollars. If they want to divide the money equally, how much will each person receive?

3. Mt. Everest is 29,028 ft. high. Mt. Whiting is 14,496 ft. high How much taller is Mt. Everest than Mt. Whitney?

4. A car traveled 330 miles in six hours. What was the average speed per hour?

5. A car's gas tank hold 22 gallons of gas. If the car can travel 31 miles per gallon, how far can the car travel on a full tank of gas?

1.
2.
3.
4.
5.
Score

Review Exercises

1. 416×22

2. 608×342

3. $500 - 76 =$

4. $405 \div 30$ (30 ⟌405)

5. $7{,}123 \div 60$ (60 ⟌7,123)

6. $9{,}963 \div 70$ (70 ⟌9,963)

Helpful Hints

Look for key words:

1. "Total," "sum," "altogether," and "in all" usually mean addition or multiplication.
2. "How much more than," "difference," "what was the increase," and "how much was left" usually mean subtraction.
3. "Divide" and "average" usually mean division.

S1. A rope is 346 feet long. If it is divided into pieces two feet long, how many pieces will there be?

S2. There are 195 students going on a field trip. If buses can hold 45 passengers, how many buses will be needed?

1. Four friends went out to dinner and the bill came to $108. If they wanted to split the bill evenly, how much will each pay?

2. A tank holds 55,000 gallons. If 33,500 gallons were removed, how many gallons would be left in the tank?

3. Sixteen bales of hay weigh 2,000 pounds. What is the weight of each bale?

4. A factory can produce 55 cars per day. How many cars can it produce in 30 days?

5. A plane traveled 3,150 miles in seven hours. What was the average speed per hour?

1.
2.
3.
4.
5.
Score

Review Exercises

1. 30 ⟌44 2. 30 ⟌532 3. 30 ⟌7,132

4. 22 ⟌45 5. 22 ⟌709 6. 22 ⟌6,931

Helpful Hints

Remember:
1. Read the problem carefully.
2. Find the important facts and numbers.
3. Decide what operation to use.
4. Solve the problem and label it with a word or short phrase.

*Always label your answer with a word or short phrase.

1.
2.
3.
4.
5.
Score

S1. A man bought a used car for $2,045. After repainting it and making several repairs, he resold it for $2,750. What was his profit?

S2. Mr. Simmons bought a new dining room set for 936 dollars. If he pays for the set in 3 equal installments, how much is each payment?

1. In an election John received 2,693 votes. Kelly received 3,109 votes. How many more votes did Kelly receive than John?

2. A field in the shape of a triangle has sides 260 feet, 317 feet, and 289 feet. How many feet is it around the field?

3. A school is buying 45 books for $1,215. What is the price per book?

4. A farmer has 2,364 eggs. If he packs them in containers that hold 12 eggs, how many containers does he need?

5. A school has thirty classes. Each class contains 28 students. What is the total number of students in the school?

Review Exercises

1. $\begin{array}{r} 236 \\ 17 \\ +\ 379 \\ \hline \end{array}$

2. $\begin{array}{r} 7{,}132 \\ -\ 1{,}427 \\ \hline \end{array}$

3. $73 + 96 + 74 + 62 =$

4. $500 - 376 =$

5. $\begin{array}{r} 426 \\ \times\ \ 7 \\ \hline \end{array}$

6. $\begin{array}{r} 307 \\ \times\ 22 \\ \hline \end{array}$

Helpful Hints

* Sometimes it is necessary to read a problem more than once.
* Sometimes it is good to circle important words and numbers.
*Sometimes drawing a sketch or diagram is useful.

1.
2.
3.
4.
5.
Score

S1. A square field has sides 258 ft. How far is it all the way around the field?

S2. Frank earns 16 dollars per hour. How much does he earn during a 40-hour work week?

1. Fremont's population is 202,156. Redford's population is 191,216. How much greater is Fremont's population than Redford's?

2. A farmer has 360 cows and decides to sell ½ of them. How many cows will he sell?

3. In a school there are 346 sixth-graders, 408 seventh-graders, and 465 eighth-graders. What is the total number of students?

4. A school has 600 students and needs to divide them into 25 equally sized classes. How many will be in each class?

5. Sam had 356 points on four tests. What was his average score?

Review Exercises

1. A car traveled 440 miles in 8 hours. What was its average speed per hour?

2. 624×7

3. A school has 25 classes with thirty students in each class. How many students are there altogether?

4. $467 + 29 + 736 + 27$

5. Billie earned $4,137 this month and $5,198 last month. How much more did she earn last month?

6. $7{,}115 - 678$

Helpful Hints

The following problems require two steps.

1. Read the problem carefully.
2. Find the important facts and numbers.
3. Decide what operations to use and in what order to use them.
4. Solve the problem.

Label your answer with a word or short phrase.

S1. Sol's test scores were 81, 90, and 99. What was her average score?

S2. A farmer had nine crates of potatoes that weighed 180 pounds each. He also had seven sacks of tomatoes that weighed a total of 365 pounds. What was the total weight of the potatoes and tomatoes?

1. A woman buys a car making a $3,000 down payment and then agrees to make 48 monthly payments of 360 dollars. What is the total cost of the car?

2. Last week a man worked 12 hours a day for five days. This week he worked 57 hours. How many hours did he work in all?

3. A theater has 15 rows of seats with 25 seats in each row. If 19 of the seats are vacant, how many are taken?

4. A tank holds 250,000 gallons of fuel. If 27,600 gallons are removed one day, and 35,750 gallons are removed the next day, how many gallons are left?

5. A car can travel 32 miles per gallon of gas. How many gallons are required to travel 448 miles? If gas is $3 per gallon, how much would the gas cost?

1.
2.
3.
4.
5.
Score

Review Exercises

1. $\begin{array}{r} 642 \\ \times\ 23 \\ \hline \end{array}$

2. A library has 47,109 fiction books and 39,783 non-fiction books. What is the total number of books in the library?

3. $30\overline{)705}$

4. If a plane travels 455 miles per hour, how far will it travel in twelve hours?

5. $12\overline{)1,976}$

6. Steve earned $65,650 this year and $58,955 last year. How much more did he earn this year?

Helpful Hints

* Sometimes reading the problem more than once is helpful.
* Finding the important facts and numbers and circling them can be helpful.
* Decide which operation comes first.
* Solve the problem. Label the answer with a word or short phrase.

S1. Sue, Harry, and Linda had a car wash, earning $375 on Friday, $360 on Saturday, and $420 on Sunday. To divide the earning equally, how much would each of them receive?

S2. Jackie wants to send out 112 party invitations. If the invitations come in boxes of 15, how many boxes must she buy? How many invitations will be left over?

1. Nine buses hold 85 passengers each. If 693 people buy tickets for a trip, how many seats won't be taken?

2. Six people are at a party. For refreshments there are three pizzas which are sliced into eight pieces each. How many pieces will each person get?

3. A carpenter bought three hammers for 17 dollars each and two saws for 23 dollars each. What was the total cost?

4. A school has 314 boys and 310 girls. If they are to be grouped into equal classes of 26 each, how many classes will there be?

5. Jim worked 48 weeks last year. Each week he worked 38 hours. If he worked an additional 240 hours of overtime, how many hours did he work in all?

1.
2.
3.
4.
5.
Score

Review Exercises

1. Joel made 540 gift cards. He wants to sell them in boxes of 30. How many boxes will he need?

2. $\begin{array}{r} 9{,}176 \\ -\ 2{,}977 \\ \hline \end{array}$

3. A garden is in the shape of a square. If each side is 29 ft., how far is it all around the garden?

4. $\begin{array}{r} 208 \\ \times\ 35 \\ \hline \end{array}$

5. Erica's test scores were 96, 88, 97, and 95. What was the total of all her scores? What was her average score?

6. $7\overline{)2{,}375}$

Helpful Hints

* Read the problem carefully.
* Find and circle the important facts and numbers.
* Decide on the correct order of operations.
* After you have solved the problem and labeled it, ask yourself whether the answer makes sense.
* Sometimes it is helpful to make a sketch.

S1. John bought a shirt for $19 and a pair of shoes for $23. If he paid with a $50 bill, how much change would he receive?

S2. A group of hikers set out on an 80-mile hike. If they hiked 9 miles per day for seven days, how many miles would be left to hike?

1. A garden in the shape of a rectangle is 140 feet long and 95 feet wide. What is the distance around the garden?

2. An orchard has eight rows of trees with 15 trees in each row. If each tree produces six bushels of fruit, how many bushels will be produced in all?

3. Bill worked 56 hours this week and 44 hours last week. If he is paid twelve dollars per hour, what were his total earnings?

4. Buses hold 65 people. If 95 parents and 165 students are attending a football game, how many buses will be needed?

5. How old will a man born in 1929 be in 2016?

1.
2.
3.
4.
5.
Score

Review Exercises

1. $41\overline{)861}$

2. The enrollment at Wilbur High School is 3,015. Last year the enrollment was 2,965. How much was this year's increase in enrollment?

3. 765 + 97 + 399 =

4. If five pounds of beef cost $65, what is the price per pound?

5. $\begin{array}{r} 605 \\ \times\ 70 \\ \hline \end{array}$

6. Yuri scored a total of 665 points on seven tests. What was her average score?

Helpful Hints

Use what you have learned to solve the following problems.

* Review some of the helpful hints from previous pages.

1.
2.
3.
4.
5.
Score

S1. Three friends took a camping trip. Food costs were 171 dollars and camping supplies were 273 dollars. If they divided the costs equally, how much would each pay?

S2. A man bought a car for $15,000. If the down payment was $3,000 and the rest was to be paid in 48 equal monthly payments, how much are the payments?

1. A school has 180 boys and 160 girls. Tickets to the class picnic are 8 dollars each. If everyone attends, what will be the cost of the tickets?

2. A new car costs $22,000. After one year it lost $2,800 of its value. After two years it lost another $3,350 in value. How much was the car's value after two years?

3. A family took a five-day trip. The first day they drove 385 miles. Each of the next four days they drove 275 miles. How many miles long was the trip?

4. A tank contained 55,000 of fuel. One day 12,500 gallons were added. The next day 15,575 gallons were removed. How many gallons are left in the tank?

5. A family bought a home for $235,000. They spent $75,500 on remodeling it. Then they sold the home for $400,000. What was the profit?

Review Exercises

1. Tom had test scores of 75, 84, and 81. What was his average score?

2. $\begin{array}{r} 423 \\ \times\ \ 27 \\ \hline \end{array}$

3. A school has twelve classes of 32 students each. How many students are there in the school?

4. $15 \overline{)3{,}015}$

5. Last year Jose worked 48 weeks and worked forty hours per week. How many hours did he work last year?

6. 9,001 − 279 =

Helpful Hints

When working with multi-step problems it is important to read very carefully and at least twice.

1. Find the important facts and numbers.
2. Decide which operations to use and in what order.
3. Draw a diagram if necessary.

S1. Jeri bought six dozen hot dogs at three dollars per dozen and three dozen burger patties at seven dollars per dozen. How much did he spend altogether?

S2. A rectangular yard is 16 feet by 24 feet. How many feet of fencing is required to enclose it? If each 5-foot section costs $35, how much will the fencing cost?

1. Two trains leave the station in opposite directions, one at 85 miles per hour, the other at 75 miles per hour. How far apart will they be in 3 hours?

2. A factory can make a table in 130 minutes and a chair in eight minutes. How long will it take to make seven tables and 15 chairs? Express the answer in hours and minutes. (60 minutes = 1 hour)

3. A farmer had 600 pounds of apples. He gave 200 pounds to neighbors and sold one-half of the remainder at three dollars per pound. How much did he make selling the apples?

4. Tom finished a 30-mile walkathon. He collected pledges of $26 for each of the first 25 miles and $50 for each remaining mile. How much did he collect in all?

5. A carpenter bought three hammers at 17 dollars each, two saws at 27 dollars each, and a drill for 58 dollars. How much did he spend in all?

1.
2.
3.
4.
5.
Score

Reviewing Whole Number Problem Solving

1.
2.
3.
4.
5.
6.
7.
8.
9.
10.
Score

1. Lincoln High School's enrollment is 4,763 and Jefferson High School's enrollment is 4,969. What is the total enrollment for both schools?

2. A plane travels 6,600 miles in twelve hours. What is its average speed per hour?

3. Sophia had test scores of 75, 83, 78, 93, and 96. What was her average score?

4. A car traveled 265 miles each day for seven days and 325 miles on the eighth day. How many miles did it travel in all?

5. A car traveled 288 miles. It averaged 24 miles per gallon of gas. If gas costs $3 per gallon, how much did the trip cost?

6. A plumber purchased 3 sinks for $129 each, three bathtubs for $372 each, and five faucets for $23 each. How much did he spend in all?

7. A school has 240 seventh-graders and 300 eighth-graders. If they are to be placed in equally sized classes of 30 students each, how many classes will there be?

8. A woman bought a car for $15,000. She made a $3,000 down payment and paid the rest in equal payments of $400. How many payments did she make?

9. A man wants to put a fence around a square lot with sides 68 feet. How many feet of fencing is needed? If each 8-foot section of fence costs $25, how much will the fence cost?

10. A family is planning a seven-day vacation. Lodging costs 125 dollars per day, food is 85 dollars per day, and entertainment is 150 dollars per day. How much will the trip cost?

Problem Solving — **Reviewing Addition and Subtraction of Fractions**

To add or subtract fractions with unlike denominators, find the least common denominator. Multiply each fraction by one to make equivalent fractions. Finally, add or subtract.

Examples:

$$\frac{2}{5} \times \frac{2}{2} = \frac{4}{10}$$
$$+\ \frac{1}{2} \times \frac{5}{5} = \frac{5}{10}$$
$$\frac{9}{10}$$

$$\frac{5}{6} \times \frac{2}{2} = \frac{10}{12}$$
$$+\ \frac{1}{4} \times \frac{3}{3} = \frac{3}{12}$$
$$\frac{13}{12} = 1\frac{1}{12}$$

When adding mixed numerals with unlike denominators, first add the fractions. If there is an improper fraction, make it a mixed numeral. Finally, add the sum to the sum of the whole numbers.

*Reduce fractions to lowest terms.

Example:

$$3\frac{2}{3} \times \frac{2}{2} = \frac{4}{6}$$
$$+\ 2\frac{1}{2} \times \frac{3}{3} = \frac{3}{6}$$
$$5 \quad \frac{7}{6} = 1\frac{1}{6} = 6\frac{1}{6}$$

To subtract mixed numerators with unlike denominators, first subtract the fractions. If the fractions cannot be subtracted, take one from the whole number, increase the fraction, then subtract.

Examples:

$$\overset{5}{\cancel{6}}\frac{1}{6} = \frac{2}{12} + \frac{12}{12} = \frac{14}{12}$$
$$-\ 3\frac{1}{4} = \frac{3}{12}$$
$$2\frac{11}{12}$$

$$7\frac{1}{2} \times \frac{3}{3} = \frac{3}{6}$$
$$-\ 2\frac{1}{3} \times \frac{2}{2} = \frac{1}{6}$$
$$5 \quad \frac{2}{6} = 5\frac{1}{3}$$

S1. $3\frac{1}{7} - 1\frac{5}{7}$

S2. $6\frac{1}{2} + 3\frac{3}{4}$

1. $\frac{8}{9} - \frac{1}{2}$

2. $\frac{8}{9} - \frac{1}{6}$

3. $7 - 2\frac{3}{5}$

4. $5\frac{1}{8} + 3\frac{1}{2}$

5. $7\frac{7}{8} + 3\frac{3}{8}$

6. $7\frac{1}{2} - 2\frac{3}{4}$

7. $3\frac{4}{5} + 4\frac{2}{3}$

8. $6\frac{1}{2} - 3$

9. $\frac{7}{16} + \frac{1}{4}$

10. $4\frac{5}{6} + 3\frac{3}{4}$

1.
2.
3.
4.
5.
6.
7.
8.
9.
10.
Score

Problem Solving

Reviewing Multiplication and Division of Fractions

When multiplying common fractions, first multiply the numerators. Next, multiply the denominators. If the answer is an improper fraction, change it to a mixed numeral.

Examples:

$\frac{3}{4} \times \frac{2}{7} = \frac{6}{28} = \boxed{\frac{3}{14}}$ $\quad$ $\frac{3}{2} \times \frac{7}{8} = \frac{21}{16} = \boxed{1\frac{5}{16}}$

If the numerator of one fraction and the denominator of another have a common factor, they can be divided out before you multiply the fractions.

Examples:

$\frac{3}{\cancel{4}_1} \times \frac{\cancel{8}^2}{11} = \boxed{\frac{6}{11}}$ $\quad$ $\frac{7}{\cancel{8}_4} \times \frac{\cancel{6}^3}{5} = \frac{21}{20} = \boxed{1\frac{1}{20}}$

When multiplying whole numbers and fractions, write the whole number as a fraction and then multiply.

Examples:

$\frac{2}{3} \times 15 =$ $\quad$ $\frac{3}{4} \times 9 =$

$\frac{2}{\cancel{3}_1} \times \frac{\cancel{15}^5}{1} = \frac{10}{1} = \boxed{10}$ $\quad$ $\frac{3}{4} \times \frac{9}{1} = \frac{27}{4} = \boxed{6\frac{3}{4}}$

To multiply mixed numerals, first change them to improper fractions, then multiply. Express answers in lowest terms.

Example: $1\frac{1}{2} \times 1\frac{5}{6} = \frac{\cancel{3}^1}{2} \times \frac{11}{\cancel{6}_2} = \frac{11}{4} = \boxed{2\frac{3}{4}}$

To divide fractions, find the reciprocal of the second number, then multiply the fractions.

Examples:

$\frac{2}{3} \div \frac{1}{2} =$ $\quad$ $2\frac{1}{2} \div 1\frac{1}{2} = \frac{5}{2} \div \frac{3}{2} =$

$\frac{2}{3} \times \frac{1}{2} = \frac{4}{3} = \boxed{1\frac{1}{3}}$ $\quad$ $\frac{5}{2} \times \frac{2}{3} = \frac{5}{3} = \boxed{1\frac{2}{3}}$

S1. $1\frac{1}{4} \times 2\frac{2}{5} =$

S2. $5\frac{1}{2} \div 1\frac{1}{2} =$

1. $\frac{12}{13} \times \frac{3}{24} =$

2. $\frac{3}{4} \times 36 =$

3. $\frac{7}{8} \times 2\frac{1}{7} =$

4. $2\frac{1}{3} \times 3\frac{1}{2} =$

5. $\frac{3}{4} \div \frac{1}{2} =$

6. $3\frac{1}{2} \div \frac{1}{2} =$

7. $3\frac{2}{3} \div 1\frac{1}{2} =$

8. $3\frac{3}{4} \div 1\frac{1}{8} =$

9. $6 \div 2\frac{1}{3} =$

10. $2\frac{2}{3} \div 2 =$

1.
2.
3.
4.
5.
6.
7.
8.
9.
10.
Score

Review Exercises

1. $\frac{1}{2} + \frac{1}{3}$

2. $1\frac{1}{5} + 2\frac{2}{3}$

3. $2\frac{1}{2} + 2\frac{2}{3}$

4. $3\frac{1}{2} - 1\frac{1}{3}$

5. $6\frac{1}{3} - 1\frac{1}{5}$

6. $7 - 1\frac{1}{5}$

Helpful Hints

1. Read the problem carefully.
2. Find and circle the important facts and numbers.
3. Decide what operations to use.
4. Solve the problem and label the answer with a word or phrase.

S1. A baker used $2\frac{1}{4}$ cups of flour for a cake and $3\frac{1}{2}$ cups for a pie. How much flour did he use in all?

S2. Steve earned 60 dollars and spend two-thirds of it. How much did he spend? (Hint: "of" means multiply.)

1. Bill weighed $124\frac{1}{4}$ pounds last year. This year he weighs $132\frac{3}{4}$ pounds. How many pounds did he gain?

2. A ribbon is $5\frac{1}{2}$ feet long. It was cut into pieces $\frac{1}{2}$ foot long. How many pieces were there?

3. It is $2\frac{1}{2}$ miles around a race track. How far will a car travel in 12 laps?

4. If it takes a man $12\frac{1}{2}$ minutes to drive to work and $16\frac{3}{4}$ minutes to drive home, what is his total commute time?

5. What is the perimeter of a square flower bed with sides $8\frac{1}{2}$ feet?

1.
2.
3.
4.
5.
Score

Review Exercises

1. $\frac{1}{2} \times \frac{4}{5} =$

2. $\frac{1}{3} \times 2\frac{1}{2} =$

3. $2\frac{1}{2} \times 1\frac{1}{5} =$

4. $\frac{2}{3} \div \frac{1}{2} =$

5. $4\frac{1}{2} \div \frac{1}{2} =$

6. $3\frac{1}{2} \div 2 =$

Helpful Hints

Look for key words:

1. "Total," "sum," "altogether," and "in all" usually mean addition or multiplication.
2. "How much more than," "difference," "what was the increase," and "how much was left" usually mean subtraction.
3. "Of" means multiply. "Divide" and "average" usually mean division.

S1. Tony bought 6 melons, each of which weighed $2\frac{1}{2}$ pounds. What was the total weight of the melons?

S2. If a cook had $5\frac{1}{4}$ pounds of beef and used $3\frac{3}{4}$ pounds, how many pounds were left?

1. Suzette worked $7\frac{1}{2}$ hours on Monday and $6\frac{3}{4}$ hours on Tuesday. How many hours did she work in all?

2. If a car travels 50 miles per hour, how far will it travel in $2\frac{1}{2}$ hours?

3. A factory can produce a tire in $2\frac{1}{2}$ minutes. How many tires can it produce in 40 minutes?

4. The Jones' had 36 pounds of beef in their freezer. They used $\frac{3}{4}$ of it. How many pounds of beef did they use?

5. Paul decided to study 12 hours for a test. If he has already studied for $7\frac{1}{4}$ hours, how much longer will he study for the test?

1.
2.
3.
4.
5.
Score

Review Exercises

1. $\frac{3}{5} + \frac{1}{2}$

2. $\frac{7}{8} - \frac{1}{4}$

3. $2\frac{2}{3} + 3\frac{1}{2}$

4. $2\frac{1}{3} + 3\frac{4}{5}$

5. $6\frac{3}{5} - 1\frac{1}{2}$

6. $6\frac{1}{5} - 2\frac{1}{2}$

Helpful Hints

***Remember:**

1. Read the problem carefully.
2. Find the important facts and numbers.
3. Decide what operation to use.
4. Solve and label your answer with a word or phrase.

* Reduce all fractions to lowest terms.

S1. Marc worked for $6\frac{1}{2}$ hours. If he is paid 12 dollars per hour, how much did he earn?

S2. A grocer has 30 pounds of tomatoes. If he wants to pack them into $2\frac{1}{2}$- pound packages, how many packages of tomatoes will he have?

1. Mike worked $3\frac{3}{4}$ hours on Monday and $2\frac{1}{2}$ hours on Tuesday. How many more hours did he work on Monday than on Tuesday?

2. Allie was $60\frac{3}{4}$ inches tall. She grew $2\frac{1}{2}$ inches. How tall is she now?

3. Phil's reading assignment is $50\frac{1}{2}$ pages. If he has read $15\frac{1}{3}$ pages, how much more does he have to read?

4. Elena bought 40 hotdogs for the picnic. If each hotdog weighed $\frac{1}{3}$ of a pound, what was the total weight of the hot dogs?

5. A family is taking a 360-mile trip. They have driven $\frac{3}{4}$ of the distance. How much farther do they have to drive?

1.
2.
3.
4.
5.
Score

Review Exercises

1. $\frac{2}{5} \times \frac{10}{11} =$

2. $5 \times 2\frac{1}{5} =$

3. $2\frac{1}{2} \times \frac{4}{5} =$

4. $\frac{3}{4} \div \frac{1}{2} =$

5. $2 \div 1\frac{1}{2} =$

6. $2\frac{2}{3} \div 1\frac{1}{3} =$

Helpful Hints

* Sometimes it's necessary to read a problem more than once.
* It's helpful sometimes to circle important words and numbers.
* Drawing a diagram can sometimes be useful.

S1. Each side of a square garden is $8\frac{1}{4}$ ft.
What is the distance around the garden?

S2. One loaf of bread weighs $1\frac{1}{2}$ pounds.
How much do 20 loaves weigh?

1. The distance around a square-shaped window is $20\frac{4}{5}$ feet.
What is the length of each side of the window?

2. Mr. Gilbert bought $5\frac{1}{2}$ pound of beef. If he divides it into 11 equally sized steaks, how much will each steak weigh?

3. Ronnie is taking a $12\frac{1}{2}$ mile hike. If he has gone $8\frac{2}{3}$ miles, how much farther does he need to hike?

4. A garden in the shape of a rectangle is $8\frac{1}{2}$ feet by $5\frac{1}{4}$ feet.
What is the total distance around the garden?

5. Remy earned $6,300 last month. If $\frac{1}{3}$ of this amount goes towards his house payment, how much is his house payment?

1.
2.
3.
4.
5.
Score

Review Exercises

1. Bill is $60\frac{1}{2}$ inches tall and Sally is $62\frac{1}{4}$ inches tall. How much taller is Sally than Bill?

2. $\frac{3}{4} + \frac{1}{3}$

3. Bill has read $\frac{3}{4}$ of a 28 page assignment. How many pages has he read?

4. $\frac{7}{8} - \frac{1}{4}$

5. Vica caught two fish. One weighed $2\frac{3}{4}$ pounds and the other weighed $3\frac{1}{2}$ pounds. What was the total weight of the fish?

6. $2\frac{3}{4} + 2\frac{1}{2}$

Helpful Hints

When working with two-step problems it is necessary to read the problems more carefully.
* Decide which operations to use and in which order.
* Reduce all fractions to lowest terms.

1.
2.
3.
4.
5.
Score

S1. A tailor had $8\frac{1}{2}$ yards of cloth. He cut off three pieces that were $1\frac{1}{2}$ yards long each. How much of the cloth was left?

S2. A man had 56 dollars. He gave his son $\frac{1}{4}$ of it and his daughter $\frac{1}{2}$ of it. How much did he have left?

1. A painter needs seven gallons of paint. He already has $2\frac{1}{2}$ gallons in one bucket, and $3\frac{1}{4}$ gallons in another. How many more gallons of paint does he need?

2. There are 30 people in a class. If $\frac{2}{5}$ of them are boys, then how many girls are in the class?

3. Eva worked $5\frac{1}{2}$ hours on Saturday and $6\frac{3}{4}$ hours on Sunday. If she was paid 12 dollars per hour, how much did she earn?

4. Michelle made 36 bracelets last week and 28 bracelets this week. If she sold $\frac{3}{8}$ of them, how many bracelets did she sell?

5. Joe's ranch has 4,000 acres. If $\frac{1}{4}$ of the ranch was used for crops and $\frac{2}{3}$ of the remainder was used for grazing, how many acres were for grazing?

Review Exercises

1. $3\frac{1}{2} \div \frac{1}{4} =$

2. How many $\frac{1}{2}$ pound beef patties are there in $5\frac{1}{2}$ pounds of beef?

3. $5 \div 1\frac{1}{2} =$

4. It takes $1\frac{1}{4}$ yards of cloth to cover a chain. How much cloth is needed to cover 8 chairs.

5. $3\frac{3}{4} \div 1\frac{1}{4} =$

6. It took Lola $12\frac{3}{4}$ minutes to drive to work and $10\frac{1}{3}$ minutes to drive home. What was her total commute time?

Helpful Hints

* Sometimes reading the problem more than once is helpful.
* Finding the important facts and numbers and circling them can be helpful.
* Carefully decide the order of operations.
* After solving the problem, the answer should make sense.

S1. A family was going to take a 360 mile trip. They travelled $\frac{1}{2}$ of the total distance the first day and $\frac{1}{3}$ of the total distance the second day. How many more miles were left to travel?

S2. Twenty feet of copper wire is cut into pieces $2\frac{1}{2}$ feet long. If each piece sells for $6, how much would all the pieces cost?

1. A farmer picked $6\frac{1}{2}$ bushels each day for five days. He then sold $15\frac{1}{2}$ bushels. How many bushels were left to sell?

2. A man bought $12\frac{3}{4}$ pounds of beef. He put $10\frac{1}{4}$ pounds in the freezer He used $\frac{3}{5}$ of the rest for cooking a stew. How many pounds did he use for the stew?

3. Land is selling for $30,000 per acre. Mr. Roberts bought $1\frac{1}{2}$ acres on Monday. On Tuesday he decided to buy an additional $2\frac{1}{2}$ acres. How much did he pay in all?

4. Mr. Jones earned $4,000. If $\frac{1}{4}$ of his pay goes towards his car payment and $\frac{2}{5}$ of his pay goes towards his house payment, what is the total cost of his housing and car payment?

5. Stewart needs to read 45 pages for English, 60 pages for history, and 63 pages for science. If he has read $\frac{2}{3}$ of the total amount of pages, how many pages has he read?

1.
2.
3.
4.
5.
Score

Review Exercises

1. $\frac{3}{4}$ of $2\frac{1}{2}$ =

2. Ellie earned 60 dollars and deposited $\frac{4}{5}$ of it into her savings account. How much did she deposit?

3. $\begin{array}{r} 736 \\ 47 \\ -\ 516 \\ \hline \end{array}$

4. A car traveled at the speed of 65 miles per hour for 8 hours. How far did the car travel?

5. $30\overline{)639}$

6. Will's weight was $155\frac{1}{4}$ pounds if he increased his weight by $5\frac{2}{3}$ pounds. How much does he weigh now?

Helpful Hints

* Read the problem carefully
* Find and circle the important facts and numbers.
* Decide on the correct order of operations.
* After you have solved the problem and labeled it, ask yourself if the answer makes sense.
* Sometimes drawing a diagram can be helpful.

S1. Al spent $\frac{1}{2}$ of his earnings and deposited $\frac{1}{3}$ of it into his savings account. What fraction of his earnings did he have left?

S2. John needs to read a 256-page novel. He must also read $\frac{3}{5}$ of a 250-page science book. What is the total number of pages that John must read?

1. Lisa rode her bike $6\frac{1}{2}$ miles on Monday, $7\frac{1}{4}$ miles on Tuesday, and $5\frac{3}{4}$ miles on Wednesday. If she wants to ride a total of 25 miles, how much farther does she need to ride?

2. Gwen has 30 ounces of cookie dough. She is going to bake $\frac{3}{4}$-ounce cookies and sell them for 2 dollars each. How many cookies will she sell, and how much will she earn selling them?

3. The Smiths had an eight-pound meatloaf and finished $\frac{3}{4}$ of it. The next day Mr. Smith finished $\frac{1}{4}$ of the leftovers. How many pounds of the original meatloaf was left?

4. Carlos needs to work 40 hours this week. He worked $6\frac{1}{2}$ hours on Monday and $5\frac{3}{4}$ hours on Tuesday. How many more hours does he need to work this week?

5. Sue wants to put trim around a square window with sides $7\frac{1}{2}$ inches. If trim comes in three-inch sections and each section costs two dollars, how much will it cost to trim the window?

1.
2.
3.
4.
5.
Score

Review Exercises

1. $2\frac{1}{2} + 3\frac{3}{4}$

2. $4\frac{1}{5} - 3\frac{2}{3}$

3. $\frac{3}{4}$ of $2\frac{1}{4} =$

4. $5 \times 1\frac{3}{4} =$

5. $\frac{3}{4} \div \frac{1}{3} =$

6. $2\frac{1}{2} \div 1\frac{1}{3} =$

Helpful Hints

Use what you have learned to solve the following problems.
* Review some of the "Helpful Hints" sections from previous pages.

S1. A student bought two binders for $2\frac{1}{2}$ dollars each and four books for $3\frac{1}{2}$ dollars each. What was the total cost?

S2. Henry spent 20 dollars for three books. The first book cost $4\frac{1}{2}$ dollars and the second book cost $5\frac{3}{4}$ dollars. What was the cost of the third book?

1. Ellen has 33 ounces of candy. She is going to put them into $2\frac{3}{4}$ ounce bags and sell them for $3 each. What will her earnings be if she sells all the candy?

2. Simi has $15\frac{3}{4}$ pounds of nuts. She is going to keep $7\frac{1}{4}$ pounds and sell the rest for six dollars a pound. How much will she receive for the nuts?

3. Tom had 60 pounds of vegetables. He sold $\frac{3}{4}$ of them and gave $\frac{2}{5}$ of the remainder to friends. How many pounds did he give to friends?

4. Mr. Jones had 40 dollars. He gave $\frac{3}{5}$ of the money to his three children for their allowances. If they divided the money equally, how much would each child receive?

5. A baker can produce $7\frac{1}{2}$ cakes in an hour. If he works for eight hours and sells each cake for 14 dollars, how much money will he make?

1.
2.
3.
4.
5.
Score

Review Exercises

1. $3\frac{3}{4} \div \frac{1}{4} =$

2. $70 \overline{)1{,}976}$

3. $5 \times 1\frac{1}{2} =$

4. $\begin{array}{r} 427 \\ \times\ 26 \\ \hline \end{array}$

5. $\frac{9}{10} - \frac{2}{3} =$

6. $\begin{array}{r} 7{,}106 \\ -\ 1{,}697 \\ \hline \end{array}$

Helpful Hints

When working with multi-step problems, remember to read the problem carefully at least twice. Also, remember your basic steps:

1. Fine the important facts and numbers
2. Decide what operations to use and in what order.
3. Solve the problem and label the answer.
4. The answer should make sense.

1.
2.
3.
4.
5.
Score

S1. Mr. Jones' class has 32 students and Mrs. Jensen's class has 40 students. If $\frac{1}{4}$ of Mr. Jones' students got A's and $\frac{3}{8}$ of Mrs. Jensen's students got A's, how many students got A's?

S2. A developer has two plots of land consisting of 39 acres and 49 acres. How many $\frac{1}{4}$-acre lots does he have? If each lot sold for $5,000, how much did he sell all the lots for?

1. There are 400 boys and 350 girls at Anderson School. Three-quarters of the boys take the bus and $\frac{3}{5}$ of the girls take the bus. How many students in all take the bus?

2. A farm has 2,000 acres. If $\frac{3}{5}$ of the farm was used for crops and $\frac{3}{4}$ of the remainder was used for grazing, how many acres were left?

3. Mr. Rosa earned $3,000 last month. $\frac{1}{3}$ of this went towards his car payment, and $\frac{3}{5}$ of the remainder was for his house payment. How much of his money was left.

4. A man had 200 pounds of beef. He gave $\frac{1}{4}$ of it away to neighbors and kept $\frac{3}{5}$ of the remainder. He sold the rest for $4 per pound. How much did he receive for the sale of the beef?

5. Julio has $7\frac{1}{2}$ acres of land. He then divides the land into $\frac{1}{4}$-acre lots and sells each of them for $5,000. If his original cost for the land was $90,000, what was his profit?

Reviewing Fractions Problem Solving

1. John earned 80 dollars and spent $\frac{3}{5}$ of it. How much did he spend?

2. Sylvia weighed $120\frac{1}{2}$ pounds. If she reduced her weight by $3\frac{3}{4}$ pounds, how much does she weigh now?

3. A carpenter had a board that was 20 feet long. If he cut it into $2\frac{1}{2}$-foot pieces, how many pieces would he have?

4. A flight to Chicago took $5\frac{2}{3}$ hours. The return flight only took $4\frac{3}{4}$ hours. What was the total time for both flights?

5. Ben worked $5\frac{1}{2}$ hours on Saturday and $6\frac{3}{4}$ hours on Sunday. If he is paid 12 dollars per hour, what was his pay?

6. Julie earned 45 dollars last week and 96 dollars this week. If she spent $\frac{1}{3}$ of her earnings, how much did she spend?

7. Mary has 30 pounds of candy that she will divide into $1\frac{1}{2}$-pound packages. How many packages will she have? If they sell for $5 each, how much money will she make?

8. In a school with 400 students, $\frac{3}{8}$ of the students take French and $\frac{2}{5}$ of the remainder take Spanish. How many take French? How many take Spanish?

9. Jean baked a pie. If she gave $\frac{1}{2}$ to Steve and $\frac{1}{3}$ to Gina, what fraction of the pie did she have left?

10. Roy picked $7\frac{3}{4}$ bushels of apples Monday and $6\frac{3}{4}$ bushels on Tuesday. If he sold $\frac{1}{2}$ of the apples for $100 per bushel, how much did he earn?

1.
2.
3.
4.
5.
6.
7.
8.
9.
10.
Score

Problem Solving — Reviewing Addition and Subtraction of Decimals

To add decimals, line up the decimal points and add as you would whole numbers. Write the decimal points in the answer. Zeroes may be placed to the right of the decimal.

Example: Add 3.16 + 2.4 + 12

```
   3.16
   1.63
+ 12.00
-------
 (17.56)
```

To subtract decimals, line up the decimal points and subtract as you would whole numbers. Write the decimal points in the answer. Zeroes may be placed to the right of the decimal.

Examples:

```
             2 11 1                  6 9 1
3.2 - 1.66 =  3.20      7 - 1.63 =   7.00
            - 1.66                 - 1.63
            (1.54)                 (5.37)
```

* Line up the decimals. * Put decimals in the answer. * Zeroes may be added to the right of the decimal.

S1.
```
  3.61
 14.4
+  .37
```

S2.
```
  7.16
- 3.473
```

1.
```
  7.16
  8.92
+ 7.634
```

2.
```
  7.6
- 1.43
```

3. 4.36 + 5.7 + 6.24 =

4. 17.2 − 8.96 =

5. 15 − 12.92 =

6. 6.93 + 5 + 7.63 =

7. .9 + .7 + .6 =

8. 7.16 − 2.673 =

9. 27.16 − 16.764 =

10. 7.73 + 2.6 + .37 + 15 =

1.
2.
3.
4.
5.
6.
7.
8.
9.
10.
Score

Problem Solving — Reviewing Multiplication and Division of Decimals

Multiplying a Decimal by a Whole Number
Multiply as you would with whole numbers.
Find the number of decimal places and place the decimal point properly in the product.

Examples:

$$\begin{array}{r} 2.32 \leftarrow \text{2 places} \\ \times \quad 6 \\ \hline \boxed{13.92} \leftarrow \text{2 places} \end{array} \qquad \begin{array}{r} 7.6 \leftarrow \text{1 places} \\ \times \ 23 \\ \hline 228 \\ 1520 \\ \hline \boxed{174.8} \leftarrow \text{1 places} \end{array}$$

Multiplying a Decimal by a Decimal
Multiply as you would with whole numbers.
Find the number of decimal places and place the decimal point properly in the product.

Examples:

$$\begin{array}{r} 2.63 \leftarrow \text{2 places} \\ \times \ .3 \leftarrow \text{1 place} \\ \hline \boxed{.789} \leftarrow \text{3 places} \end{array} \qquad \begin{array}{r} .724 \leftarrow \text{3 places} \\ \times \ .23 \leftarrow \text{2 places} \\ \hline 2172 \\ 14480 \\ \hline \boxed{.16652} \leftarrow \text{5 places} \end{array}$$

Dividing a Decimal by a Whole Number
Divide as you would with whole numbers.
Place the decimal point directly up.

Examples:

$$\begin{array}{r} 2.8 \\ 3\overline{)8.4} \\ -6\downarrow \\ \hline 24 \\ -24 \\ \hline 0 \end{array} \qquad \begin{array}{r} .084 \\ 3\overline{).252} \\ -24\downarrow \\ \hline 12 \\ -12 \\ \hline 0 \end{array}$$

Dividing a Decimal by a Decimal
Move the decimal point in the divisor the number of places necessary to make it a whole number. Move the decimal point in the dividend the same number of places.

Examples:

$$\begin{array}{r} .8 \\ .3.\overline{).2.4} \\ -2\,4 \\ \hline 0 \end{array} \qquad \begin{array}{r} 950.^{*} \\ .03.\overline{)28.50.} \\ -27\downarrow \\ \hline 15 \\ 15 \\ \hline 0 \end{array}$$

*Sometimes placeholders are necessary.

S1. $\begin{array}{r} 3.24 \\ \times \ 2.3 \\ \hline \end{array}$

S2. $.5\overline{).325}$

1. $\begin{array}{r} 4.26 \\ \times \quad 3 \\ \hline \end{array}$

2. $\begin{array}{r} 3.4 \\ \times \ 16 \\ \hline \end{array}$

3. $\begin{array}{r} 3.08 \\ \times \ 1.6 \\ \hline \end{array}$

4. $\begin{array}{r} 6.32 \\ \times \ 23.4 \\ \hline \end{array}$

5. $2\overline{)2.68}$

6. $5\overline{)7.3}$

7. $.003\overline{)1.2}$

8. $.5\overline{).375}$

9. $.15\overline{).0075}$

10. $8.7\overline{).1131}$

1.
2.
3.
4.
5.
6.
7.
8.
9.
10.
Score

Review Exercises

1. $\begin{array}{r} 3.56 \\ 2.723 \\ +\ 4.96 \\ \hline \end{array}$

2. $3.16 + 15 + 2.79 =$

3. $7.13 - 2.652 =$

4. $\begin{array}{r} 6.2 \\ -\ 3.564 \\ \hline \end{array}$

5. $\begin{array}{r} 2.14 \\ \times\ \ 3 \\ \hline \end{array}$

6. $\begin{array}{r} .58 \\ \times\ 32 \\ \hline \end{array}$

Helpful Hints

1. Read the problem carefully.
2. Find the important facts and numbers.
3. Decide which operation to use.
4. Solve the problem and label it with a work or short phrase.

* Be careful with decimal placement.

S1. If calculators cost $7.95 each, what is the cost of eight calculators?

S2. A jet traveled 1,250 miles in 2.5 hours. What was its average speed?

1. Potatoes cost $3.23 a pound and carrots cost $2.89 a pound. How much more do potatoes cost per pound?

2. A man bought a desk for $375.50, a chair for $119.90, and a lamp for $23.45. What was the total cost of the items?

3. A square lot has sides 48.5 feet long. How far is it around the lot?

4. 12 cans of corn cost $13.68. What is the cost of one can?

5. A baseball glove was on sale for $32.65. If the regular price was $45.25, how much can be saved buying it on sale?

1.
2.
3.
4.
5.
Score

Review Exercises

1. $.72 \times .3$

2. $2.62 \times .03$

3. $.003 \times .002$

4. $2 \overline{)3.68}$

5. $5 \overline{).13}$

6. $2 \overline{).014}$

Helpful Hints

Look for key words:

1. "Total," "sum," "altogether," and "in all" usually mean addition or multiplication.
2. "How much more than," "difference," "what was the increase," and "how much was left" usually mean subtraction.
3. "Divide" and "average" usually mean division.

1.
2.
3.
4.
5.
Score

S1. A plane can travel 450 miles in one hour. At this rate, how far can it travel in .8 hours?

S2. A sack of potatoes was $3.15. If the price was $.45 per pound, how many pounds were in the sack?

1. Tom weighs 135.6 pounds and Jerry weighs 142.75 pounds. What is their total weight?

2. If 8 pounds of butter costs $7.12, what is the price per pound?

3. An engine uses 3.5 gallons of gas per hour. How many gallons will the engine use in 3.2 hours?

4. If cans of soda cost $.25, how many cans of soda can be bought with $5.00?

5. Steak costs $4.80 per pound. How much will .7 pounds cost?

Review Exercises

1. $.2 \overline{)1.34}$

2. $.02 \overline{).13}$

3. $.03 \overline{)5.1}$

4. $.06 \overline{).324}$

5. $.12 \overline{)1.104}$

6. $.18 \overline{).576}$

Helpful Hints

Sometimes it is necessary to read a problem at least twice.
Then decide the necessary operation and solve the problem.

* Be careful with decimal placement.
* Think about your answer and be sure it makes sense.

S1. A man bought groceries that cost a total of $34.16. If he paid with a 50-dollar bill, how much change did he receive?

S2. John bought six boxes of chocolates for $7.95 each. What was the total cost?

1. Twelve pounds of apples cost $27.00. What is the price per pound?

2. Lonnie bought a car for $6545.25 and resold it for $8476.49. What was his profit?

3. Four friends earned $301.92. If they decided to divide the money equally, how much will each person receive?

4. Ken bought a car for $3006.00. If he pays for it in twelve equal payments, how much is each payment?

5. How much will it cost for twenty-four $.41 stamps?

1.
2.
3.
4.
5.
Score

Review Exercises

1. $\begin{array}{r} 7.36 \\ .95 \\ +\ 4.93 \\ \hline \end{array}$

2. 6 – 2.713 =

3. 6 x .427 =

4. $\begin{array}{r} \$2.75 \\ \text{x}\quad 8 \\ \hline \end{array}$

5. $5\overline{)\$15.75}$

6. $.02\overline{)5}$

Helpful Hints	
	* When reading a problem, sometimes it is helpful to circle important facts and numbers. * When the problem has been solved, go back and be sure the answer makes sense. * Sometimes drawing a diagram is helpful.

S1. Samuel earns $15.50 per hour. How much does he earn in 40 hours?

S2. A rectangle has a length of 12.5 inches and a width of 8.3 inches. What is the perimeter of the rectangle?

1. A man has $481.24 in his savings account. If he makes a deposit of $242.35, what will his new balance be?

2. The sale price of a dress is $14.50 less than the regular price. If the regular price is $49.65, what is the sale price?

3. Rainfall for the last four days was 2.12 inches, 3.89 inches, 2.73 inches, and 4.79 inches. What was the total rainfall?

4. A man bought 12.5 gallons of gas for $40.00. What was the price per gallon?

5. 1.5 pounds of beef cost $5.10. What is the price per pound?

1.
2.
3.
4.
5.
Score

Review Exercises

1. $5\overline{)3}$

2. Brian earned 40 dollars and spent $\frac{3}{4}$ of it. How much did he spend?

3. $\frac{3}{5} + \frac{1}{3}$

4. A $7\frac{1}{2}$ foot board is cut into 3 equal pieces. How long is each piece?

5. $3\frac{1}{7} - 1\frac{1}{2}$

6. What is the perimeter of a square-shaped window with sides $7\frac{1}{2}$ inches?

Helpful Hints

When working with 2-step problems it is necessary to read the problems more carefully.
* Decide which operations to use and in which order.
* Be carefully with decimal placement.
* Be sure your answer makes sense.

S1. A man bought five bags of chips at $.89 each and a pizza for $8.95. How much did he spend?

S2. Yuri bought a hammer for $6.79 and a screwdriver for $4.75. If he paid with a 20-dollar bill, what was his change?

1. Zach is taking a trip of 192 miles. If his car gets 24 miles per gallon of gas, and gas costs $3.10 per gallon, what is the cost of the trip?

2. Mark worked 40 hours and was paid $12.50 per hour. He received a bonus of $125.75 for overtime. What were his total earnings?

3. Cans of corn are two for $1.19. What is the cost for 12 cans?

4. Jeans are on sale for $12.95. If the regular price is $15.50, how much would be saved by buying two pairs of jeans on sale?

5. Three friends earned $3.65 on Monday, $7.75 on Tuesday, and $9.75 on Wednesday. If they divided the money equally, how much would each of them receive?

1.
2.
3.
4.
5.
Score

Review Exercises

1. A car traveled 365 miles per day for five days. What was the total distance traveled.

2. $2 \div 1\frac{3}{4} =$

3. A factory can produce 34 engines per hour. How many can it produce in 12 hours?

4. $4 \times 2\frac{1}{3} =$

5. A rope $17\frac{1}{2}$ feet long is divided into 7 equal pieces. How long is each piece?

6. $2\frac{1}{2} \div 1\frac{1}{4} =$

Helpful Hints

Ask yourself these questions:
1. What does the problem ask you to do?
2. Are there helpful "keywords"?
3. What are the facts and numbers that will help in finding the answer?

S1. A jacket costs $12.00. Sales tax is .07 of this price. What will be the total price including sales tax? (Hint: "of" often means multiply.)

S2. A man bought 12 gallons of gas at $2.75 per gallon and a quart of oil for $4.79. What was the total cost?

1. Cans of peas are 3 for $1.29. How much would one can of peas and a bag of chips priced at $2.19 cost altogether?

2. Tom's times for the 100-yard dash were 11.8, 12.2, and 12.3. What was his average time?

3. If two pounds of chicken cost $3.60, how much will ten pounds cost?

4. John bought a shirt for $13.55 and a pair of shoes for $27.50. If he gave the clerk a $50 bill, how much change will he receive?

5. Marie is buying a bike. She makes a down payment of $75.00 and pays the rest in 12 monthly installments of $32.75 each. How much does she pay in all?

1.
2.
3.
4.
5.
Score

Review Exercises

1. A car traveled 265 miles in five hours. What was its average speed?

2. $3\overline{)7{,}968}$

3. Last year's attendance at a concert was 13,768. This year the attendance was 17,272. What was the increase in attendance?

4. $\begin{array}{r} 307 \\ \times\ 26 \\ \hline \end{array}$

5. It is $1\frac{1}{2}$ miles around a track. What is the distance traveled in 10 laps?

6. 376 + 39 + 778 =

Helpful Hints

Remember these important steps:

1. Read the problem carefully.
2. Find the important facts and numbers.
3. Decide what operations are necessary and the order in which to use them.
4. Solve the problem and label it with a word or short phrase.

S1. A number is the sum of 12.6 and 14.78, decreased by 2.36. Find the number.

S2. A number is the product of 6.2 and 3.4, minus 1.76. Find the number.

1. A number is the quotient of 15.05 and 7, plus 3.16. Find the number.

2. A number is the difference of 17.2 and 9.16, multiplied by 2.1. Find the number.

3. Tony worked 8 hours per day for 6 days. If he was paid \$10.50 per hour, what were his earnings?

4. A group of five friends go to a restaurant. The bill comes to \$65.75 plus a tip of \$9.00. If they decide to split the cost evenly, how much will each of them pay?

5. Bill bought 2.5 pounds of beef priced at \$3.50 per pound. What will be his change from a \$10.00 bill?

1.
2.
3.
4.
5.
Score

Review Exercises

1. 3.26 + 4 + 3.96 =
2. 7 − 2.165 =
3. 3 x 7.096 =
4. 5 ⟌ 1.3
5. .3 ⟌ .015
6. .003 ⟌ 15

Helpful Hints

Use what you have learned to solve the following problems.
* "Of" often means multiply.
* Sometimes drawing a diagram is helpful.

S1. A runner ran 7.8 miles on Monday, 8.4 miles on Tuesday, and 7.5 miles on Wednesday. What was his average distance?

S2. A rancher decided to sell .2 of his 4,000 acre ranch. If he sold the land for $7,000 per acre, how much was the sale price?

1. The regular price of a pen is $4.75 and the sale price is $3.85. How much can be saved buying 50 pens on sale?

2. Beef is $4.50 per pound and chicken is $2.50 per pound. What is the cost of .6 pounds of beef and .8 pounds of chicken?

3. Alex worked 30 hours, earning $8.15 per hour. If he bought a video game for $97.50, how much of his earnings was left?

4. Jane wants to trim a painting with a length of 9,5 inches and a width of 7.5 inches. If trim costs $5.00 per inch, what will be the cost?

5. Tom bought 6 CD's for $5.50 each and a DVD for $11.79. What was the total cost?

1.
2.
3.
4.
5.
Score

Review Exercises

1. $2\frac{3}{4} \div \frac{1}{2} =$

2. $2\frac{3}{4} \times \frac{1}{2} =$

3. $7\frac{1}{4} - 3\frac{2}{3}$

4. $5\frac{1}{5} + 3\frac{7}{8}$

5. $7 - 2\frac{1}{4}$

6. $7\frac{1}{2} - 2\frac{3}{4}$

Helpful Hints

When working with multi-step problems, remember to read the problem carefully at least twice to fully understand what is being asked.
* Circling key words and numbers can be quite helpful.
* Make sure the answer makes sense. * "Of" often means multiply.

1.
2.
3.
4.
5.
Score

S1. A woman bought 6 dozen hotdogs at \$3.39 per dozen and three dozen burger patties at \$6.15 per dozen. How much did she spend in all?

S2. A rectangular yard is 24 feet by 16 feet. How many feet of fence is needed to enclose it? If each 2.5-foot section costs \$30.00, how much will the fencing cost?

1. Mr. Arnold's class has 40 students and Ms. Grey's class has 35 students. If .6 of Mr. Arnold's students got A's and .4 of Ms. Grey's students got A's, how many students got A's altogether?

2. A carpenter bought 3 hammers for \$7.99 each, 2 saws for \$14.50 each, and a drill for \$26.95. What was the total cost?

3. A farm had 2,000 acres. If .6 of the land was used for crops and .4 of the remainder was used for grazing, how many acres were left?

4. Two cars leave a garage in opposite directions, one at 50.5 miles per hour and the other at 30.25 miles per hour. How far apart will they be in 2 hours?

5. Beans are three cans for \$.69 and corn is three cans for \$1.23. How much will it cost for two cans of each?

Reviewing Decimal Problem Solving

1. A plane can travel 560 miles in one hour. At this rate, how far can it travel in .8 hours?

2. If seven pounds of butter cost $8.33, what is the price per pound?

3. Arnie weighed 139.5 pounds last year, and this year he weighs 147.3 pounds. How much did his weight increase?

4. A woman earned $127.50 on Monday, $133.79 on Tuesday, and $127.65 on Wednesday. What were her total earnings?

5. If potatoes cost $.55 per pound, how many pounds can be bought with $4.95?

6. On a 308-mile trip, a car averaged 22 miles for each gallon of gas. If gas costs $3.15 per gallon, how much did the trip cost?

7. A student bought five pens for $2.19 each and a binder for $8.39. What was the total cost?

8. Beef is $3.60 per pound and chicken is $3.20 per pound. What is the total cost for 2.5 pounds of beef and 1.5 pounds of chicken?

9. A ranch had 5,000 acres. The owner sold .3 acres of the ranch for $7,000 per acre. How much did he receive for the sale of the land?

10. A woman bought three chairs for $21.95 each and a table for $26.50. How much change did she receive if she paid with a $100.00 bill?

1.
2.
3.
4.
5.
6.
7.
8.
9.
10.
Score

Percent means "per hundred" or "hundredths." If a fractions is expressed as hundredths, it can easily be written as a percent.

Examples: $\frac{7}{100} = 7\%$ $\frac{3}{10} = 30\%$ $\frac{19}{100} = 19\%$

"Hundredths" = percent

Decimals can easily be changed to percents.

Examples: .27 = 27% .9 = .90 = 90%

* Move the decimal twice to the right and add a percent symbol.

To change a fraction to a percent, first change the fraction to a decimal, then change the decimal to a percent. Move the decimal twice to the right and add a percent symbol.

Examples:

$\frac{3}{4}$ 4 ⟌ 3.00 = .75 = (75%) ; - 2.8

$\frac{16}{20} = \frac{4}{5}$ 5 ⟌ 4.00 = .80 = (80%) ; - 4.0 ; 0

Percents can be expressed as decimals and as fractions. The fraction form may sometimes be reduced to its lowest terms.

Examples: $.25\% = .25 = \frac{25}{100} = \frac{1}{4}$ $8\% = .08 = \frac{8}{100} = \frac{2}{25}$

S1. 20% = . = ____ S2. 9% = . = ____

1. 16% = . = ____ 2. 6% = . = ____

3. 75% = . = ____ 4. 40% = . = ____

5. 1% = . = ____ 6. 45% = . = ____

7. 12% = . = ____ 8. 5% = . = ____

9. 50% = . = ____ 10. 13% = . = ____

1.
2.
3.
4.
5.
6.
7.
8.
9.
10.
Score

Finding the Percent of a Number
To find the percent of a number, you may use either fractions or decimals. Use what is the most convenient.

Examples:

Find 25% of 60
.25 x 60

$$\begin{array}{r} 60 \\ \times\ .25 \\ \hline 300 \\ 120 \\ \hline 15.00 \end{array}$$

OR

$$\frac{25}{100} = \frac{1}{4}$$

$$\frac{1}{4} \times \frac{60}{1} = \frac{15}{1} = 15$$

Finding the Percent When finding the percent, first write a fraction, change the fraction to a decimal, then change the decimal to a percent.

Examples:

4 is what percent of 16?

$$\frac{4}{16} = \frac{1}{4}$$

$4 \overline{)1.00} = .25 = 25\%$; -8, 20, -20, 0

5 is what percent of 25?

$$\frac{5}{25} = \frac{1}{5}$$

$5 \overline{)1.00} = .20 = 20\%$; -1.0, 00

Finding the Whole
To find the whole when the part and the percent are known, simply change the equal sign "=" to the division sign "÷".

Examples:

6 = 25% of what number
6 ÷ 25% change = to ÷
6 ÷ .25 change % to decimal

$.25 \overline{)6.00} = 24.$

12 is 80% of what?
12 ÷ 80%
12 ÷ .8

$.8 \overline{)12.0} = 15.$

* Be careful to move decimal points carefully.

Solve the problems.

S1. 4 is what % of 20

S2. 3 = 15% of what?

1. Find 20% of 210.

2. Find 6% of 350.

3. 15 is what % of 60?

4. 5 is 20% of what?

5. 15 = 75% of what?

6. 30% of 200 =

7. 18 is what % of 24?

8. Find 25% of 64.

9. 3 is 5% of what?

10. 16 is what % of 80?

1.
2.
3.
4.
5.
6.
7.
8.
9.
10.
Score

Review Exercises

1. $723 \times .6$

2. $39.7 \times .06$ (2.)

3. Find 12% of 60.

4. Find 25% of 70.

5. $\frac{3}{5} \times 25 =$

6. $\frac{1}{4} \times 20 =$

Helpful Hints

When finding the percent of a number in a word problem, you can change the percent of a fraction or a decimal. Always express your answer in a short phrase or sentence.

Example:
A team played 60 games and won 75% of them.
How many games did they win?
Find 75% of 60
.75 x 60

$$\begin{array}{r} 60 \\ \times\ .75 \\ \hline 300 \\ 420 \\ \hline 45.00 \end{array}$$

OR

$$\frac{75}{100} = \frac{3}{4}$$

$$\frac{3}{\not{4}_1} \times \frac{\not{60}^{15}}{1} = \frac{45}{1} = 45$$

Answer:
The team won 45 games.

S1. Gloria took a test with 40 problems on it. If he got 80% of the problems correct, how many problems did he get correct?

S2. If 6% of the 500 students enrolled in a school are absent, how many students are present?

1. Marty wants to buy a car that costs $9,000. If he has saved 20% of this amount, how much has he saved?

2. Erin has a stamp collection consisting of 30 stamps. If 70% of the stamps are from the USA, how many stamps are from other countries?

3. A house priced at $150,000 requires a 20% down payment. How much is the down payment?

4. A coat is priced at $60. If the sales tax is 7% of the price, how much is the sales tax? What is the total cost including sales tax?

5. If a car costs 15,000 and loses 20% of its value in one year, how much will the car be worth in a year?

1.
2.
3.
4.
5.
Score

Review Exercises

1. Find 6% of 80.
2. Find 60% of 80.
3. Change $\frac{3}{5}$ to a percent.
4. 3 is what % of 5?
5. 15 is what % of 20?
6. .9 is what % ?

Helpful Hints

Use what you have learned to solve the following problems.
* Before solving the problem, change the percent to a fraction or a decimal.

S1. A ranch is 5,000 acres. If 70% of the ranch is used for crops, how many acres are used for other purposes?

S2. A computer priced at $350 is on sale for 15% off. What is the sale price of the computer?

1. A school has 300 students. If 40% of the students are boys, how many girls are there in the school?

2. Gordon bought a CD for $16. If the sales tax is 8%, what it the total cost of the CD?

3. Bill had a 90-page reading assignment. If he decided to read 60% of the pages before he eats dinner, how many pages will he have left to read after dinner?

4. A book was priced at $7.50. If the price was reduced by 20%, what is the new price?

5. A team played 40 games and won 80% of them. How many games did they win?

1.
2.
3.
4.
5.
Score

Review Exercises

1. Change $\frac{2}{5}$ to a percent.
2. Change $\frac{21}{28}$ to a percent.
3. 5 is what % of 20?
4. 18 is what % of 20?
5. Find 30% of 80.
6. Find 12% of 80.

Helpful Hints

When finding the percent first write a fraction, change the fraction to a decimal, then change the decimal to a percent.

Example:
A team played 20 games and won 15 of them.
What percent of the games did they win?

15 is what % of 20?

$$\frac{15}{20} = \frac{3}{4}$$

$$\begin{array}{r} .75 \\ 4\overline{)3.00} \\ -2\,8 \\ \hline 20 \\ -20 \\ \hline 0 \end{array} \quad .75 = 75\%$$

Answer:
They won 75% of the games.

S1. On a test with 25 questions, Felicia got 20 correct. What percent did she get correct?

S2. In a class of 25 students, 15 are girls. What percent are boys?

1. A worker earned $200 and spent $150 of the money. What percent of the earnings did he spend?

2. If 21/25 of a class were present at school, what percent of the class was present?

3. A pitcher threw 12 pitches and nine of them were strikes. What percent were strikes?

4. On a test with 28 questions, Anna got 21 of them correct. What percent of the questions did she miss?

5. Twelve of the 20 students in a class rode the bus to school. How many students ride the bus?

1.
2.
3.
4.
5.
Score

Review Exercises

1. Find 6% of 90.
2. 12 is what % of 15?
3. 3 = 20% of what?
4. Find 40% of 65.
5. Change $\frac{15}{50}$ to a percent.
6. 6 = 25% of what?

Helpful Hints

Use what you have learned to solve the following problems.

* When solving the problem be sure to reduce fractions to lowest terms.

S1. Peter has finished 18 of the 24 questions on a test. What percent of the test has he finished?

S2. Sally earned $25. She saved $15 and spent $10. What percent of her earnings did she spend?

1. A team played 12 games this month and 13 games last month. If they won a total of 20 games, what percent of the games played did they win?

2. Forty players tried out for a team and only 12 made the team. What percent of those who tried out made the team?

3. Twenty-seven is what percent of 36?

4. Jill earned $60. She spent $15 on a calculator and $9 on a binder. What percent of her earnings did she spend?

5. Three-fifths of a class received A's. What percent of the class received A's?

1.
2.
3.
4.
5.
Score

Review Exercises

1. 3 is 20% of what?
2. 40 is 25% of what?
3. Find 6% of 200.
4. Find 60% of 200.
5. 3 is what % of 15?
6. 45 is what % of 50?

Helpful Hints

When finding the whole, simply change the equal sign to division. **Examples:**

5 people got A's on a test.
This is 20% of the class.
How many are in the class?

5 = 20% of what?
$5 \div 25\%$
$5 \div .2$

$.2\overline{)5.0}$ = 25.

There are 25 in the class.

200 students at a school are 7th graders. If this is 25% of the students, how many students are there in the school?

200 = 25% of what?
$200 \div 25\%$
$200 \div .25$

$.25\overline{)200.00}$ = 800.

There are 800 students in the school.

S1. A team won five games. If this is 20% of the total games played, how many games have they played?

S2. Bill has 24 USA stamps in his collection. If this is 20% of his collection, how many stamps does he have? How many are not USA stamps?

1. A man spent six dollars, which was 20% of his earnings. How much were his earnings?

2. Seven is 20% of what number?

3. A farmer decided to sell 50 bushels of corn. If this is 5% of his corn harvest, how many bushels were in his harvest?

4. A basketball player made nine shots. This was 75% of the shots taken. How many shots did he take? How many shots did he miss?

5. Sonya got 24 problems correct on a test. Her score was 80%. How many problems were on the test?

1.
2.
3.
4.
5.
Score

Review Exercises

1. 3 is 15 % of what?
2. 2 is what percent of 5?
3. Find 90% of 250.
4. 15 is 60% of what?
5. 28 is what % of 35?
6. Find 9% of 250.

Helpful Hints

Use what you have learned to solve the following problems.
* Be careful with decimal placement.

1.
2.
3.
4.
5.
Score

S1. If you get a score of 70% and you got 35 questions correct, how many questions are on the test?

S2. Twenty people passed a test. This was 16% of those who took the test. How many took the test?

1. There are eight red marbles in a bag. If this is 40% of the marbles in the bag, how many marbles are in the bag? How many are not used?

2. Mr. Garcia paid $5,000 in taxes. If this was 20% of his earnings, how much were his earnings?

3. Six is 40% of what?

4. Fifteen students in a class received an award. How many students were in the class if this was 30% of the class?

5. A team won 12 games. If this was 75% of the games played, how many games did they lose?

Reviewing Percent Problem Solving

1. Find 25% of 220

2. 6 is 20% of what?

3. 8 is what % of 40?

4. Change $\frac{19}{20}$ to a percent.

5. In a class of 40 people, 20% received A's. How many received A's?

6. A team played 24 games and won 18 of them. What percent of the games did they win?

7. Seven students qualified to take advanced French. If this was 20% of the class, how many students are in the class?

8. If $\frac{12}{16}$ of the students in a school walk to school, what percent of the students walk to school?

9. A ranch has 6,000 acres. If 60% of the ranch is used for crops, how many acres are used for other purposes?

10. In a class of 30 students, 6 of them received A's. What percent did not receive A's?

1.
2.
3.
4.
5.
6.
7.
8.
9.
10.
Score

More Reviewing Percent Problem Solving

1. Change $\frac{12}{20}$ to a percent.

2. 6 is 15% of what number?

3. 8 is what percent of 32?

4. Find 15% of 450.

5. A team lost five games. If this was 20% of all the games they played, how many games did they play?

6. A school has 450 students. If 60% of the students take the bus to school, how many students do not take the bus to school?

7. In a school of 800 students, 600 of them plan to attend summer school. What percent are planning to attend summer school?

8. Seventy percent of the 80 fish in a tank are goldfish. How many goldfish are in the tank?

9. In a class of 40 students, five are taking French and 25 are taking Spanish. What percent of the class is taking a foreign language?

10. A CD is priced at $32. If the sales tax is 8%, what will be the total cost of the CD?

1.
2.
3.
4.
5.
6.
7.
8.
9.
10.
Score

Final Review - Whole Numbers

1. $\begin{array}{r} 347 \\ +\ 467 \\ \hline \end{array}$

2. $\begin{array}{r} 614 \\ 723 \\ 17 \\ +\ 824 \\ \hline \end{array}$

3. 6,403 + 763 + 16,799 =

4. 6,502 + 2,134 + 654 + 24 =

5. 6,093 + 748 + 83 + 769 =

6. $\begin{array}{r} 927 \\ -\ 648 \\ \hline \end{array}$

7. $\begin{array}{r} 5{,}392 \\ -\ 1{,}764 \\ \hline \end{array}$

8. 6,053 - 4,639 =

9. 5,000 - 3,286 =

10. 6,003 - 719 =

11. $\begin{array}{r} 73 \\ \times\ 4 \\ \hline \end{array}$

12. $\begin{array}{r} 7{,}136 \\ \times\ 4 \\ \hline \end{array}$

13. $\begin{array}{r} 45 \\ \times\ 37 \\ \hline \end{array}$

14. $\begin{array}{r} 342 \\ \times\ 46 \\ \hline \end{array}$

15. $\begin{array}{r} 643 \\ \times\ 246 \\ \hline \end{array}$

16. $4\overline{)526}$

17. $4\overline{)1376}$

18. $40\overline{)568}$

19. $30\overline{)7614}$

20. $18\overline{)1243}$

1.
2.
3.
4.
5.
6.
7.
8.
9.
10.
11.
12.
13.
14.
15.
16.
17.
18.
19.
20.
Score

Final Review - Fractions

1. $\frac{4}{7} + \frac{1}{7}$

2. $\frac{7}{8} + \frac{3}{8}$

3. $\frac{3}{5} + \frac{1}{3}$

4. $3\frac{1}{2} + 2\frac{3}{8}$

5. $7\frac{3}{5} + 6\frac{7}{10}$

6. $\frac{7}{8} - \frac{1}{8}$

7. $6\frac{1}{4} - 2\frac{3}{4}$

8. $5 - 2\frac{1}{7}$

9. $7\frac{3}{5} - 2\frac{1}{2}$

10. $7\frac{1}{4} - 2\frac{1}{3}$

11. $\frac{3}{5} \times \frac{2}{7} =$

12. $\frac{3}{20} \times \frac{5}{11} =$

13. $\frac{5}{6} \times 24 =$

14. $\frac{5}{8} \times 3\frac{1}{5} =$

15. $2\frac{1}{2} \times 3\frac{1}{2}$

16. $\frac{5}{6} \div \frac{1}{3} =$

17. $2\frac{1}{3} \div \frac{1}{2} =$

18. $2\frac{2}{3} \div 2 =$

19. $5\frac{1}{2} \div 1\frac{1}{2} =$

20. $6 \div 1\frac{1}{3} =$

1.
2.
3.
4.
5.
6.
7.
8.
9.
10.
11.
12.
13.
14.
15.
16.
17.
18.
19.
20.
Score

Final Review - Decimals

1. $\begin{array}{r} 4.67 \\ 3.5 \\ +\ 3.743 \\ \hline \end{array}$

2. $.6 + 7.62 + 6.3 =$

3. $16.8 + 6 + 7.9 =$

4. $\begin{array}{r} 36.4 \\ -\ 17.8 \\ \hline \end{array}$

5. $\begin{array}{r} 6.3 \\ -\ 3.69 \\ \hline \end{array}$

6. $72 - 1.68 =$

7. $\begin{array}{r} 2.64 \\ \times\ \ 3 \\ \hline \end{array}$

8. $\begin{array}{r} 2.6 \\ \times\ 73 \\ \hline \end{array}$

9. $\begin{array}{r} .63 \\ \times\ 2.4 \\ \hline \end{array}$

10. $\begin{array}{r} .126 \\ \times\ 4.23 \\ \hline \end{array}$

11. $10 \times 3.65 =$

12. $1,000 \times 3.6 =$

13. $2\overline{)4.64}$

14. $5\overline{)6.7}$

15. $.4\overline{).124}$

16. $.004\overline{)1.2}$

17. $.15\overline{).0045}$

18. $.67\overline{)8.71}$

19. Change $\frac{3}{5}$ to a decimal.

20. Change $\frac{7}{20}$ to a decimal.

1.
2.
3.
4.
5.
6.
7.
8.
9.
10.
11.
12.
13.
14.
15.
16.
17.
18.
19.
20.
Score

Final Review - Graphs

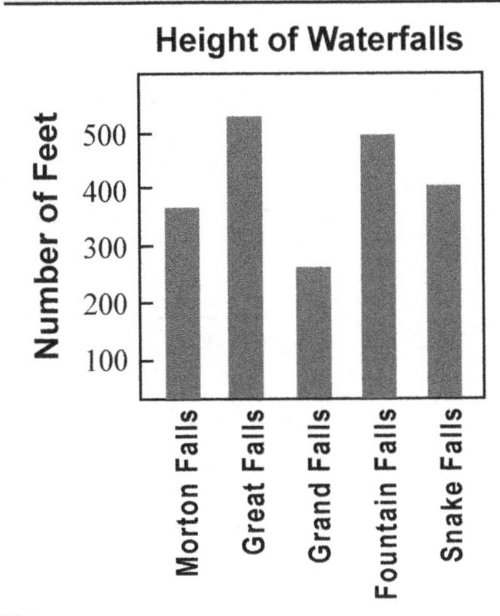

1. Which waterfall is the smallest?
2. Approximately how high is Great Falls?
3. Approximately how much higher is Morton Falls than Grand Falls?
4. Which waterfall is about the same height as Morton Falls?
5. Which waterfall is the fourth highest?

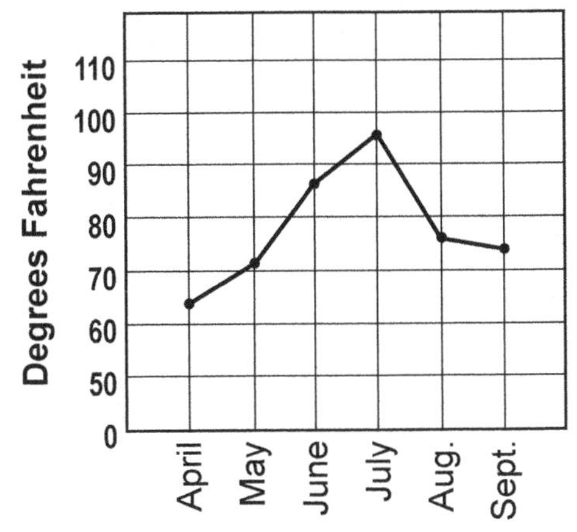

11. What is the average temperature of May?
12. How much cooler was April than July?
13. Which was the second hottest month?
14. Which month's temperature dropped the most from the previous month?
15. What is the difference in temperature between the hottest month and the second hottest month?

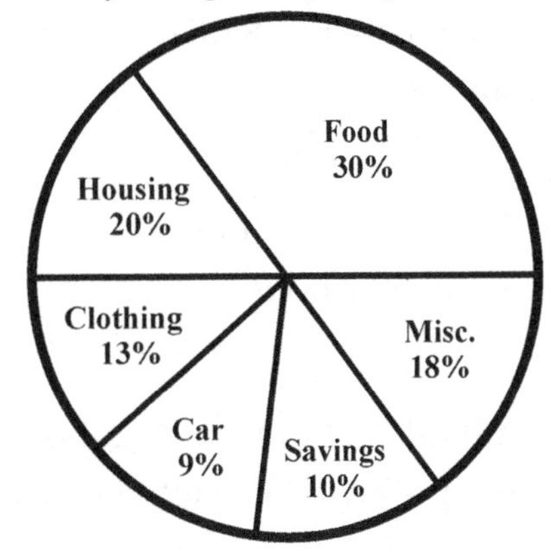

6. What percent of the family's money is spent on clothing?
7. What percent of the budget is spent on housing?
8. How many dollars are spent on housing per month? (Hint: Find 20% of $3,000.)
9. How many dollars do they save per month?
10. What percent of the family budget is left after paying food, housing, and car expenses?

Fish caught in Drakes Bay in 1991	
Salmon	
Perch	
Cod	
Bass	
Snapper	
Tuna	
Each symbol represents 10,000 fish	

16. How many perch were caught in 1991?
17. How many more snapper were caught than bass?
18. How many salmon and cod were caught?
19. If the average perch weighs three pounds, how many pounds of perch were caught in 1991?
20. What are the three most commonly caught types of fish?

1.
2.
3.
4.
5.
6.
7.
8.
9.
10.
11.
12.
13.
14.
15.
16.
17.
18.
19.
20.
Score

Final Review - Problem Solving

1. A man earned \$1,496 last week and \$2,018 this week. What were his total earnings?

2. A plane travelled 3,015 miles in five hours. What was the average speed per hour?

3. A student took a test with 60 problems and got $\frac{5}{6}$ of them correct. How many problems did he get correct?

4. Sixteen pounds of nuts were put into bags which each held $1\frac{1}{3}$ pounds. How many bags were there?

5. Alicia earned $16\frac{1}{2}$ dollars on Monday and $12\frac{3}{4}$ dollars on Tuesday. How much more did she make on Monday?

6. If a train can travel 260 miles in one hour, how far can it travel in .8 hours?

7. Hats were on sale for \$23.50. If the regular price was \$30.25, how much would be saved buying three hats on sale?

8. Stan weighed 96 pounds, Cheryl weighed 120 pounds, and Renee weighed 93 pounds. What was their average weight?

9. Tony earned \$65.50 each day for five days and \$15.25 on the sixth day. How much did he earn in all?

10. If eight pounds of apples cost \$22.48, how much does one pound cost?

1.
2.
3.
4.
5.
6.
7.
8.
9.
10.
Score

Final Review - Problem Solving

11. Antonia took a test with 32 questions. If he missed $\frac{1}{8}$ of them, how many questions did he get correct?

12. A car travelled 348 miles and averaged 29 miles per gallon of gas. If gas was $2.50 per gallon, how much did the trip cost?

13. To tune up his car, Stan bought six spark plugs for $1.95 each and a filter for $9.95. What was the total cost?

14. Cans of corn are three for $.76. How much would 15 cans cost?

15. In a school with 320 students, 60% are boys. How many students are boys?

16. A man can pay for a car in 36 payments of $160 or pay $3,500 cash. How much can he save by paying in cash?

17. John earned 84 dollars and spent 21 dollars. What percent of his earnings did he spend?

18. Maria baked a pie and gave $\frac{1}{3}$ of it to Ellen and $\frac{2}{5}$ of it to Alfonso. What fraction of the pie did she have left?

19. Twelve people in a class passed the final exam. If this was 80% of the class, how many are in the class?

20. A man has a rectangular lot which is 36 feet by 20 feet. How many feet of fencing is needed to enclose it? If each four-foot section costs $35, how much will the fence cost?

11.
12.
13.
14.
15.
16.
17.
18.
19.
20.
Score

Final Test - Whole Numbers

1. $\begin{array}{r} 342 \\ 53 \\ +\ 616 \\ \hline \end{array}$

2. $\begin{array}{r} 746 \\ 716 \\ 823 \\ +\ 634 \\ \hline \end{array}$

3. 7,362 + 775 + 72,516

4. 7,013 + 2,615 + 776 + 29 =

5. 7,001 + 696 + 18 + 732 =

6. $\begin{array}{r} 743 \\ -\ 367 \\ \hline \end{array}$

7. $\begin{array}{r} 5{,}282 \\ -\ 1{,}367 \\ \hline \end{array}$

8. 7,052 - 2,637 =

9. 6,000 - 3,678 =

10. 7,001 - 678 =

11. $\begin{array}{r} 76 \\ \times\ 3 \\ \hline \end{array}$

12. $\begin{array}{r} 7{,}653 \\ \times\ 4 \\ \hline \end{array}$

13. $\begin{array}{r} 53 \\ \times\ 46 \\ \hline \end{array}$

14. $\begin{array}{r} 627 \\ \times\ 36 \\ \hline \end{array}$

15. $\begin{array}{r} 673 \\ \times\ 346 \\ \hline \end{array}$

16. $3\overline{)425}$

17. $6\overline{)1697}$

18. $30\overline{)769}$

19. $42\overline{)8992}$

20. $28\overline{)1577}$

1.
2.
3.
4.
5.
6.
7.
8.
9.
10.
11.
12.
13.
14.
15.
16.
17.
18.
19.
20.
Score

Final Test - Fractions

1. $\frac{3}{5} + \frac{1}{5}$

2. $\frac{5}{6} + \frac{3}{6}$

3. $\frac{2}{3} + \frac{1}{5}$

4. $3\frac{2}{3} + 4\frac{5}{9}$

5. $7\frac{3}{4} + 2\frac{3}{8}$

6. $\frac{5}{8} - \frac{1}{8}$

7. $7\frac{2}{5} - 2\frac{3}{5}$

8. $7 - 2\frac{3}{5}$

9. $6\frac{3}{4} - \frac{1}{2}$

10. $9\frac{1}{3} - 3\frac{2}{5}$

11. $\frac{2}{3} \times \frac{4}{7} =$

12. $\frac{12}{13} \times \frac{3}{24} =$

13. $\frac{3}{4} \times 36 =$

14. $\frac{7}{8} \times 2\frac{1}{7} =$

15. $2\frac{1}{3} \times 3\frac{1}{2}$

16. $\frac{3}{4} \div \frac{1}{2} =$

17. $3\frac{1}{2} \div \frac{1}{2} =$

18. $3\frac{2}{3} \div 1\frac{1}{2} =$

19. $3\frac{3}{4} \div 1\frac{1}{8} =$

20. $6 \div 2\frac{1}{3} =$

1.
2.
3.
4.
5.
6.
7.
8.
9.
10.
11.
12.
13.
14.
15.
16.
17.
18.
19.
20.
Score

Final Test - Decimals

1. $\begin{array}{r} 3.72 \\ 4.6 \\ +\ 3.963 \\ \hline \end{array}$

2. $.3 + 2.96 + 7.1 =$

3. $15.4 + 4 + 9.7 =$

4. $\begin{array}{r} 37.3 \\ -\ 16.7 \\ \hline \end{array}$

5. $\begin{array}{r} 7.1 \\ -\ 2.37 \\ \hline \end{array}$

6. $6 - 1.43 =$

7. $\begin{array}{r} 3.12 \\ \times\ \ 3 \\ \hline \end{array}$

8. $\begin{array}{r} 3.4 \\ \times\ 16 \\ \hline \end{array}$

9. $\begin{array}{r} .47 \\ \times\ 1.6 \\ \hline \end{array}$

10. $\begin{array}{r} .436 \\ \times\ 3.21 \\ \hline \end{array}$

11. $100 \times 2.36 =$

12. $1{,}000 \times 2.7 =$

13. $2\overline{)2.68}$

14. $5\overline{)7.3}$

15. $.5\overline{).325}$

16. $.003\overline{)1.2}$

17. $.15\overline{).0075}$

18. $8.7\overline{).1131}$

19. Change $\frac{7}{8}$ to a decimal.

20. Change $\frac{11}{25}$ to a decimal.

1.
2.
3.
4.
5.
6.
7.
8.
9.
10.
11.
12.
13.
14.
15.
16.
17.
18.
19.
20.
Score

Final Test - Graphs

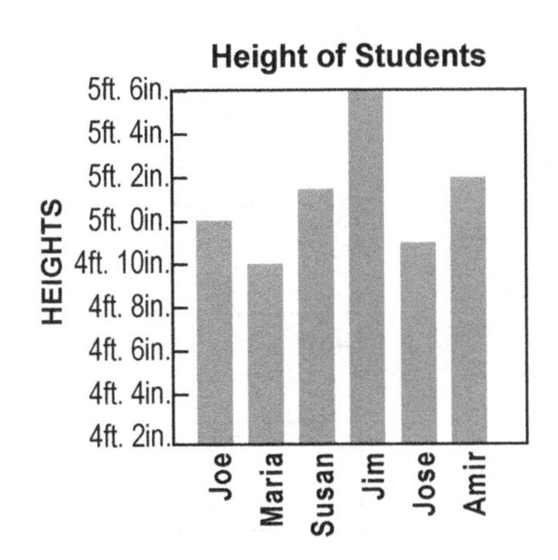

1. Which student is 5' 0"?
2. Who are the two tallest students?
3. Who is the fourth tallest student?
4. What is the combined height of Maria and Joe?
5. How many inches taller is Jim than Maria?

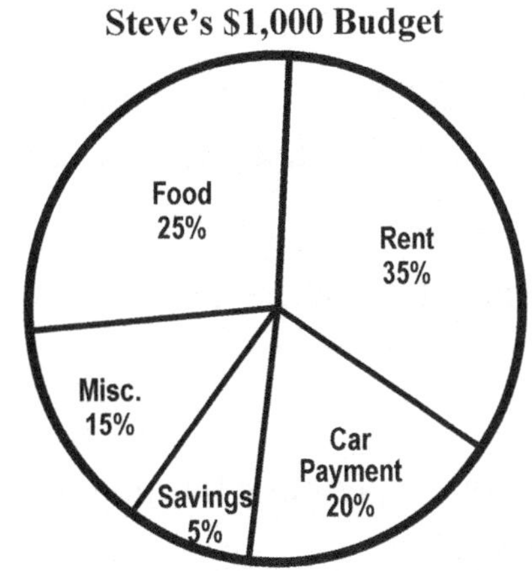

6. What percent of the budget is spent on rent?
7. What percent is spent on food?
8. How many dollars are spent on rent? (Hint: Find 35% of $1,000.)
9. How many dollars are saved per month?
10. What percent is left after food and rent?

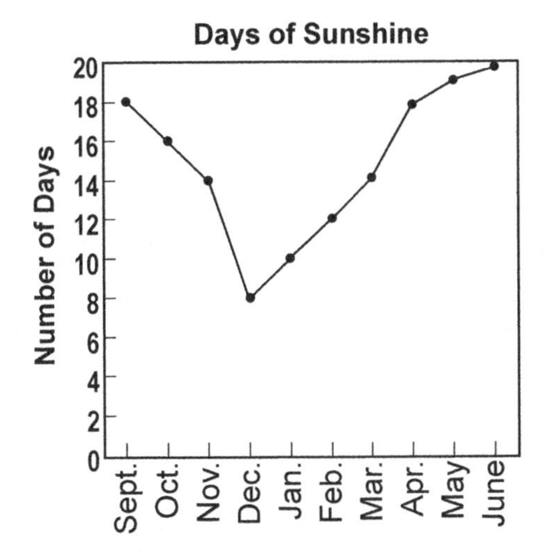

11. How many days of sunshine were there in December?
12. How many more sunny days were there in May than in March?
13. Which month had the third fewest sunny days?
14. What is the total number of sunny days in October and June?
15. Which month had the second most sunny days?

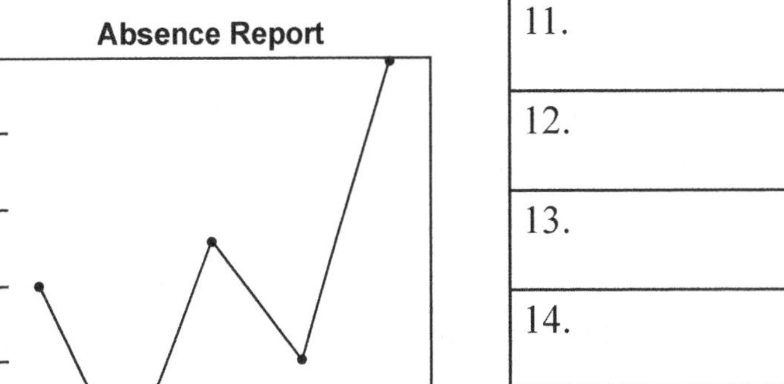

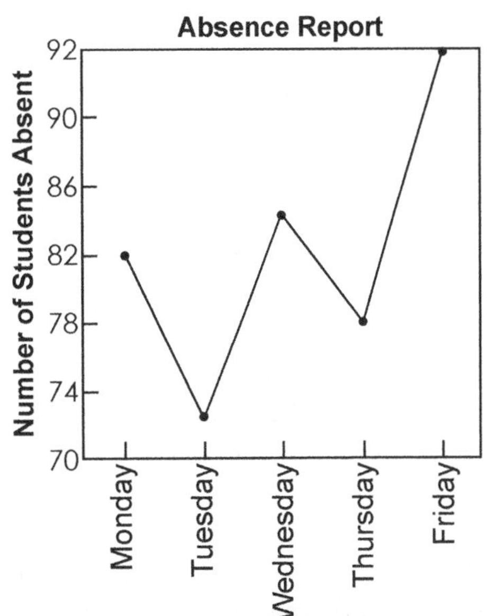

16. Which day has the most absences?
17. Which day had the most students present?
18. Which day had the fewest absences?
19. What was the total number of students absent on Monday and Thursday?
20. Which two days had the most absences?

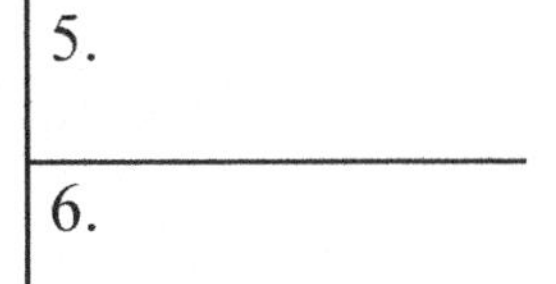

1.
2.
3.
4.
5.
6.
7.
8.
9.
10.
11.
12.
13.
14.
15.
16.
17.
18.
19.
20.

Score

Final Test - Problem Solving

1. Mt. Baxter is 12,496 feet high and Mt. Henry is 13,998 feet high. How much higher is Mt. Henry than Mr. Baxter?

2. A class has 35 students. If $\frac{2}{5}$ of them are boys, how many girls are in the class?

3. A plane traveled 485 miles per hour for nine hours. What was the total distance traveled?

4. A rope nine feet long was cut into $1\frac{1}{2}$ - foot pieces. How many pieces were there?

5. Six pounds of corn cost $4.14. What is the price for two cans?

6. A shirt cost $15. If the sales tax was 8%, what was the total cost of the shirt?

7. Manuel scored a total of 344 points on four tests. What was his average score?

8. A race trace is $2\frac{1}{2}$ miles long. How far would you travel if you drove 12 laps around the track?

9. A tailor has nine yards of cloth. He cut two pieces which were $1\frac{2}{5}$ yards each. How much cloth was left?

10. Will is taking a test with 45 problems and he has finished $\frac{3}{5}$ of them. How many problems are left?

1.
2.
3.
4.
5.
6.
7.
8.
9.
10.
Score

Final Test - Problem Solving

11. A boy bought four sodas for $.79 each and five burgers for $1.19 each. What was the total cost?

12. Four people paid $60.80 for dinner and left a 20% tip. If they divided the cost evenly, how much did each pay?

13. Three tickets to a movie cost $9.75. How much will twelve tickets cost?

14. A man bought a car for $24,000. If he made a $4,000 down payment and paid the rest in equal payments of $500, how many payments would he make?

15. A rancher had 500 pounds of beef. He gave $\frac{2}{5}$ of it away and put $\frac{1}{2}$ of the remainder in his freezer. He sold the rest for $4.00 per pound. How much did he receive for the sale of the beef?

16. A team played 15 games and won 12 of them. What percent of the games did the team lose?

17. A school has 350 boys and 320 girls. If $\frac{3}{5}$ of the boys take French and $\frac{5}{8}$ of the girls take French, how many students are taking French?

18. Ferdie bought a CD for $12.00 and a DVD for $15.00. What is the total cost if sales tax is 8%?

19. Cans of soda are three for $1.68. How much would two sodas and a bag of chips priced at $1.79 cost altogether?

20. Dean took a test with 20 problems and got 19 of them correct. What percent did he get correct?

11.
12.
13.
14.
15.
16.
17.
18.
19.
20.
Score

Section 3

Pre-Algebra

Review Exercises

Note to students and teachers: This section will include necessary review problems from all types covered in this book. Here are some sample problems with which to get started.

1. $364 + 79 + 716 =$ 2. $705 - 269 =$ 3. $7 \times 326 =$

Helpful Hints

A set is a well-defined collection of objects. A = {1,2,3,4,5} is read, "A is the set whose members are 1, 2, 3, 4, and 5. Each object in a set is called an element or member. **Infinite sets** are sets whose number of members is uncountable. **Example: A = {1,2,3...}** **Finite sets** are sets whose number of members is countable. **Example: B = {3,4,5}** **Disjoint sets** have no members in common. The **null set** or **empty set** is the set with no members, and is written as { } or Ø. **Equivalent sets** can be paired in a one-to-one correspondence. **Example: A = {1,2,3,4}**

B = {2,3,4,5}

U = the universal set; the set that contains all the members.

∈ means "is a member of."

∉ means "is not a member of."

Use the information and examples given in the Helpful Hints to answer the following questions. Explain each answer in the space below.

S1. Is A = {2,4,6,8,10} an infinite set?

S2. Is B = {2,4,6...} a finite set?

1. Are A = {1,2,3} and B = {3,4,5} disjoint sets?
2. Are C = {0,1,2,3} and D = {2,4,6,8} equivalent sets?
3. List two disjoint sets.
4. List two equivalent sets.

For 5-10, list the members of each set.

5. {the odd numbers between 2 and 12}
6. {the even numbers less than 13}
7. {the whole numbers between 2 and 10}
8. {the multiples of five between 9 and 32}
9. The members common to A = {1,2,3,4,5} and B = {1,3,5,7}
10. {the whole numbers greater than 7 and less than 13}

1.
2.
3.
4.
5.
6.
7.
8.
9.
10.
Score

Problem Solving

In a class of 38 students, one-half are girls. How many girls are there in the class?

Review Exercises

1. List two disjoint sets.
2. List two equivalent sets.
3. List an infinite set.
4. List a finite set.
5. Are A = {1,5,10} and B = {5,10,15} disjoint sets? Why?
6. Are C = {2,4,5} and D = {0,1,2,3} equivalent sets? Why?

Helpful Hints

Use the sets below for the following examples that pertain to subset, intersection, and union.

A = {1,2,3} B = {0,1,2,3,4} C = {2,4,6,8,10}

If **A** and **B** are sets and all the members of **A** are members of **B**, then **A** is a **subset** of **B** and is written **A ⊂ B**. **Example:** Is **A ⊂ B**? Yes, because all the members of **A** are members of **B**.

If **A** and **B** are sets then **A intersection B** is the set whose members are included in both sets **A** and **B**, and is written **A ∩ B**. **Example:** Find **A ∩ C**

A ∩ C = {2} (Two is the only member included in both **A** and **C**.)

If A and B are sets then **A union B** is the set whose members are included in **A** or **B**, or both **A** and **B**, and is written **A ∪ B**. **Example:** Find **B ∪ C**

B ∪ C = {0,1,2,3,4,6,8,10} (**B ∪ C** contains all members in **B**, **C**, or both **B** and **C**.)

Use the sets below to answer the questions on this page.
Explain in the space if necessary.

A = {5,6,7} B = {1,2,3,4,5,6,7} C = {1,2,4,5,7,8} D = {1,2,4,6,8,10}

S1. Is A ⊂ B? Why?

S2. Find A ∩ B.

1. Find B ∪ C.
2. Is A ⊂ D? Why?
3. List all subsets of A.
   (Hint: there are seven of them.)
4. Find B ∩ C.
5. Find C ∩ D.
6. Find A ∪ B.
7. Find B ∪ D.
8. Find B ∩ D.
9. Are C and D equivalent sets? Why?
10. Are A and D disjoint sets? Why?

1.
2.
3.
4.
5.
6.
7.
8.
9.
10.
Score

Problem Solving

Three weeks ago you Jose sold seven of his baseball cards from his collection, and last week he bought 12 new cards. If he now has 85 cards, how many did he start with three weeks ago?

Review Exercises

Use A = {1,2,3,5,6}, B = {2,4,8}, and C = {1,2,3,6} to answer the following questions.

1. Find A ∩ B.
2. Find B ∪ C.
3. Find A ∩ C.
4. Find B ∩ C.
5. Are A and C equivalent sets? Why?
6. Is A an infinite set? Why?

Helpful Hints

Integers are the set of whole numbers and their opposites.

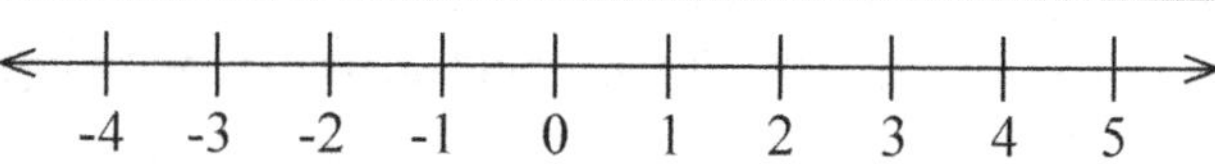

Integers to the left of zero are negative and less than zero. Integers to the right of zero are positive and greater than zero. When two integers are on a number line, the one farthest to the right is greater. Hint: When adding integers, always find the sign of the answer first.

Examples: The sum of two negatives is a negative.

-7 + -5 = - (the sign is negative)

7
+5
(-12)

When adding a negative and a positive, the sign is the same as the integer farthest from zero. Then subtract.

-7 + 9 = + (the sign is positive)

9
-7
(+2)

S1. -9 + 12 =

S2. -15 + -6 =

1. -15 + 29 =
2. -12 + -6 =
3. 42 + -56 =
4. -15 + -16 =
5. 8 + 32 =
6. -39 + 76 =
7. -96 + -72 =
8. 73 + -86 =
9. -15 + -19 =
10. 71 + -81 =

1.
2.
3.
4.
5.
6.
7.
8.
9.
10.

Score

Problem Solving

At 3:00 a.m. the temperature was -8°. By 6:00 a.m. the temperature was another -12° colder. What was the temperature at 6:00 a.m.?

Review Exercises

1. -16 + 9 =
2. -6 + 19 =
3. -26 + -13 =
4. -26 + 26 =
5. Carefully define "set."
6. Carefully define "finite set."

Helpful Hints

When adding more than two integers, group the negatives and positives separately, then add.

Examples:

-6 + 4 + -5 =
-11 + 4 = -
(sign is negative)

$$\begin{array}{r} 11 \\ -4 \\ \hline 7 \end{array} = \boxed{-7}$$

7 + -3 + 8 + 6 =
-11 + 13 = +
(sign is positive)

$$\begin{array}{r} 13 \\ -11 \\ \hline 2 \end{array} = \boxed{+2}$$

S1. -3 + 5 + -6 =

S2. -7 + 6 + -9 + 3 =

1. -3 + -4 + 5 =
2. 7 + -6 + -8 =
3. -15 + 19 + -12 =
4. -6 + 9 + 7 + 4 =
5. -16 + 32 + -18 =
6. -13 + 16 + -8 + 15 =
7. -9 + -7 + -6 =
8. -3 + 7 + -8 + -9 =
9. -32 + 16 + -17 + 8 =
10. -76 + 25 + -33 =

1.
2.
3.
4.
5.
6.
7.
8.
9.
10.
Score

Problem Solving

Alice started the week with no money. On Monday she earned \$45.00. On Tuesday she spent \$27.00. On Wednesday she earned \$63.00. On Thursday she spent \$26.00. How much money does she have left?

Review Exercises

Use the sets to answer problems 1 through 6.

A = {1,4,8,9,12}, B = {0,5,10,15}, and C = {9,10,11,15} to answer the following questions.

1. Find A ∩ B.
2. Find A ∪ B.
3. Find B ∩ C.
4. Find A ∪ C.
5. Find A ∪ Ø.
6. Find B ∩ Ø.

Helpful Hints

To subtract an integer means to add to its opposite.

Examples:

-3 - -8 =
-3 + 8 = +
(sign is positive) 8 - 3 = 5 = (+5)

8 - 10 =
8 + -10 = -
(sign is negative) 10 - 8 = 2 = (-2)

6 - -7 =
6 + 7 = +
(sign is positive) 7 + 6 = 13 = (+13)

S1. -6 - 8 =

S2. 6 - 9 =

1. 3 - -9 =
2. 15 - 18 =
3. -16 - -25 =
4. -16 - 12 =
5. 32 - -14 =
6. -35 - 14 =
7. -6 - 4 =
8. -64 - -53 =
9. -49 - 54 =
10. -63 - -78 =

1.
2.
3.
4.
5.
6.
7.
8.
9.
10.
Score

Problem Solving

A boy jumped off a diving board that was 15 feet high. He touched the bottom of the pool that was 12 feet below the surface of the water. How far is it from the diving board to the bottom of the pool?

Review Exercises

1. $-72 + 16 =$
2. $55 + -33 =$
3. $-16 + -19 =$
4. $7 - 16 =$
5. $-5 - 6 =$
6. $-5 - -9 =$

Helpful Hints

The product of two integers with different signs is negative. The product of two integers with the same sign is positive. (• means multiply.)

Examples:

$7 \cdot -16 = -$ (sign is negative)

$$\begin{array}{r} 16 \\ \times\ 7 \\ \hline 112 \end{array} = \boxed{-112}$$

$-8 \cdot -7 = +$ (sign is positive)

$$\begin{array}{r} 8 \\ \times\ 7 \\ \hline 56 \end{array} = \boxed{+56}$$

When multiplying more than two integers, group them into pairs to simplify. An integer next to parenthesis means to multiply.

Examples:

$2 \cdot -3\ (-6) =$
$(2 \cdot -3)\ (-6) =$
$-6\ (-6) = +$
(sign is positive)

$$\begin{array}{r} 6 \\ \times\ 6 \\ \hline 36 \end{array} = \boxed{+36}$$

$-2 \cdot -3 \cdot 4 \cdot -2 =$
$(-2 \cdot -3)\ (4 \cdot -2) =$
$6 \cdot -8 = -$
(sign is negative)

$$\begin{array}{r} 6 \\ \times\ 8 \\ \hline 48 \end{array} = \boxed{-48}$$

S1. $-3 \times 16 =$

S2. $-18 \cdot 7 =$

1. $-4 \cdot -17 =$
2. $16 \times -4 =$
3. $-24 \cdot -12 =$
4. 23×-16
5. $-23 \cdot 32 =$
6. $(-2)\ (-3)\ (-4) =$
7. $-8\ (-1) \cdot 1\ (-4) =$
8. $4\ (-3) \cdot 2\ (-3) =$
9. $(-3)\ (-2)\ (3)\ (4) =$
10. $10\ (-11)\ (-3) =$

Answers
1.
2.
3.
4.
5.
6.
7.
8.
9.
10.
Score

Problem Solving

An elevator started on the 28th floor. It went up seven floors, down 13 floors, and up nine floors. On what floor is the elevator located now?

Review Exercises

1. $-27 + 16 =$
2. $-37 + -19 =$
3. $7 - 9 =$
4. $-6 - -8 =$
5. $5 \bullet -7 =$
6. $-2 \bullet -6 \bullet 3 =$

Helpful Hints

The quotient of two integers with different signs is negative. The quotient of two integers with the same signs is positive. (HINT: Determine the sign, then divide.)

Examples:

$36 \div -4 = -$ (sign is negative)

$$\begin{array}{r} 9 \\ 4\overline{)36} \\ -36 \\ \hline 0 \end{array} = \boxed{-9}$$

$\frac{-123}{-2} = +$ (sign is positive)

$$\begin{array}{r} 41 \\ 3\overline{)123} \\ -12\downarrow \\ \hline 3 \end{array} = \boxed{+41}$$

Use what you have learned to solve problems like these.

Examples:

$$\frac{-36 \div -9}{4 \div -2} = \frac{4}{-2} = \boxed{-2}$$ (sign is negative)

$$\frac{4 \times -8}{-8 \div 2} = \frac{-32}{-4} = \boxed{+8}$$ (sign is positive)

S1. $-36 \div 9 =$

S2. $\frac{-90}{-15} =$

1. $-64 \div 4 =$
2. $-336 \div -7 =$
3. $\frac{-75}{-5} =$
4. $104 \div -4 =$
5. $\frac{54 \div -9}{-18 \div -9} =$
6. $\frac{16 \div -2}{-1 \times -4} =$
7. $\frac{-75 \div -25}{-3 \div -1} =$
8. $\frac{42 \div -2}{-3 \bullet -7} =$
9. $\frac{45 \div -5}{-9 \div 3} =$
10. $\frac{-56 \div -7}{-36 \div -9} =$

1.
2.
3.
4.
5.
6.
7.
8.
9.
10.
Score

Problem Solving

At midnight the temperature was 7°. By 2:00 a.m. the temperature had dropped 12°. By 4:00 a.m. it had dropped another 6°. What was the temperature at 4:00 a.m.?

Reviewing All Integer Operations

1. $-9 + 7 =$

2. $9 + -7 =$

3. $-9 + -7 =$

4. $-7 + -8 + 14 =$

5. $-32 + 16 + 21 + -24 =$

6. $7 - 9 =$

7. $4 - -9 =$

8. $-3 - 9 =$

9. $-13 - 14 =$

10. $16 - 17 =$

11. $3 \bullet -16 =$

12. $-4 \bullet -19 =$

13. $2\,(-7)\,(-4) =$

14. $-2 \bullet 3\,(-4) \bullet 2 =$

15. $-36 \div 4 =$

16. $-126 \div -3 =$

17. $\frac{-128}{-8} =$

18. $\frac{-36 \div 2}{24 \div -4} =$

19. $\frac{6 \bullet -3}{-54 \div -6} =$

20. $\frac{20 \bullet -3}{-30 \div -10} =$

1.
2.
3.
4.
5.
6.
7.
8.
9.
10.
11.
12.
13.
14.
15.
16.
17.
18.
19.
20.

Review Exercises

Use the following sets to find the answers.

$A = \{1,3,4,5,9\}$, $B = \{1,2,4,6\}$, and $C = \{1,3,6,7\}$

1. Find $A \cap B$.
2. Find $B \cup C$.
3. Find $B \cap C$.
4. Find $A \cup B$.
5. Find $A \cup C$.
6. Find $A \cup \emptyset$.

Helpful Hints

The rules for integers apply to positive and negative fractions.

Examples:

$-\frac{1}{2} + \frac{3}{5} =$

$-\frac{5}{10} + \frac{6}{10} = +$ (the sign is positive)

$\frac{6}{10} - \frac{5}{10} = \frac{1}{10}$

$-\frac{3}{5} + \frac{1}{3} =$

$-\frac{9}{15} + -\frac{5}{15} = -$ (the sign is negative)

$\frac{9}{15} + \frac{5}{15} = -\frac{14}{15}$

$-\frac{3}{5} \times 1\frac{1}{3} = -$

A negative multiplied or divided by a positive is negative.

$\frac{3}{5} \times -\frac{3}{2} = -\frac{9}{10}$

$-\frac{2}{3} \div -\frac{1}{2}$

A negative divided by a negative is a positive.

$-\frac{2}{3} \times -\frac{2}{1} = \frac{4}{3} = 1\frac{1}{3}$

S1. $-\frac{1}{5} + \frac{1}{2} =$

S2. $\frac{1}{2} + -\frac{2}{5} =$

1. $\frac{1}{2} - \frac{3}{4} =$

2. $-\frac{2}{3} + -\frac{1}{2} =$

3. $-\frac{4}{5} \times 2\frac{1}{2} = -$

4. $\frac{5}{8} + -\frac{1}{4} =$

5. $-\frac{1}{3} - \frac{1}{4} =$

6. $-1\frac{2}{3} \times -1\frac{1}{2} =$

7. $\frac{3}{4} + \frac{1}{3} =$

8. $\frac{5}{8} + -\frac{1}{2} =$

9. $2\frac{1}{2} \div -\frac{1}{4} =$

10. $-\frac{1}{5} + \frac{2}{3} =$

1.
2.
3.
4.
5.
6.
7.
8.
9.
10.

Score

Problem Solving

There are two sixth-grade classes. One has 35 students and another has 32 students. If a total of 17 sixth graders received A's, how many did not receive A's?

Review Exercises

1. -75 + 16 =
2. -19 - 17 =
3. 16 × -4 =
4. -9 - 19 =
5. -36 ÷ -9 =
6. 2 • -7 × -2 =

Helpful Hints

The rules for integers apply to positive and negative decimals.

Example:

- .71 + .9 = + (the sign is positive)

```
  .90
- .71
  .19
```

Example:

-.5 × 1.23 = - (a negative multiplied or divided by a positive is a negative)

```
  1.23
×   .5
- .615
```

Example:

-2.9 - 3.2 =
-2.9 + - 3.2 = - (the sign is negative)

```
  2.9
+ 3.2
 -6.1
```

Example:

-3.12 ÷ - .3 = + (a negative multiplied or divided by a negative is a positive)

```
    10.4
.3 | 3.12
```

Work the following problems. If necessary, review the rules for integers.

S1. -3.21 + 2.3 =

S2. 5.15 ÷ -.5 =

1. -5.2 - 7.61 =
2. 5.63 + -2.46 =
3. -.7 × 6.12 =
4. 5.9 - -6.23
5. -7.11 ÷ -3 =
6. -.72 + .9 =
7. -2.13 × -.2 =
8. -6.2 + -.73 =
9. 5.2 + -3.19 =
10. -5.112 ÷ .3 =

1.
2.
3.
4.
5.
6.
7.
8.
9.
10.
Score

Problem Solving

Anna weighed 120.5 pounds. She lost 7.3 pounds and then gained back 4.8 pounds. How much does she weigh now?

Review Exercises

1. $-.3 + .7 =$
2. $-2.7 + -3.2 =$
3. $3 \times -2.6 =$
4. $-\frac{1}{2} + -\frac{1}{3} =$
5. $\frac{2}{5} + -\frac{1}{2} =$
6. $-\frac{1}{2} \times -1\frac{1}{5} =$

Helpful Hints

In the expression 5^3, the number 5 is called the **base** and the number 3 is called the **power** or **exponent**. The exponent tells how many times the base is to be multiplied by itself. In the example 5^3, you would multiply 5 three times: $5^3 = 5 \times 5 \times 5 = 125$. Negative numbers can have exponents: $(-2)^3 = (-2) \times (-2) \times (-2) = 4 \times (-2) = -8$. Any number to the power of 1 = the number. Any number to the power of 0 = 1.

Examples:

$3^4 = 3 \times 3 \times 3 \times 3$
$= 9 \times 9$
$= 81$

$(-5)^4 = (-5) \times (-5) \times (-5) \times (-5)$
$= 25 \times 25$
$= 625$

$5^1 = 5$
$6^0 = 1$

S1. $4^2 =$

S2. $-3^3 =$

1. $6^3 =$
2. $5^0 =$
3. $(-2)^4 =$
4. $2^5 =$
5. $7^1 =$
6. $8^3 =$
7. $(-1)^5 =$
8. $5^5 =$
9. $(-5)^3 =$
10. $(-3)^4 =$

1.
2.
3.
4.
5.
6.
7.
8.
9.
10.
Score

Problem Solving

A certain number to the third power is equal to eight. What is the number?

Review Exercises

1. $7^2 =$
2. $9^3 =$
3. $(-6)^2 =$
4. $5 + -6 + 8 + -3 =$
5. $7^0 =$
6. $9^1 =$

Helpful Hints

Many numbers can be written as exponents. **Examples:**

$5 \times 5 \times 5 \times 5 = 5^4$

$7 \times 7 \times 7 \times 7 \times 7 = 7^5$

$(-2) \times (-2) \times (-2) = (-2)^3$

$(-60) \times (-60) \times (-60) = (-60)^3$

$125 = 5^3$

$36 = 6^2$ or $(-6)^2$

$8 = 2^3$

$25 = 5^2$ or $(-5)^2$

Rewrite each of the following as an exponent.

S1. $12 \times 12 \times 12 =$

S2. $27 =$

1. $2 \times 2 \times 2 \times 2 \times 2 \times 2 =$
2. $(-9) \times (-9) \times (-9) =$
3. $16 \times 16 \times 16 \times 16 =$
4. $49 =$
5. $100 =$
6. $121 =$
7. $(-1) \times (-1) \times (-1) \times (-1) =$
8. $32 =$
9. $16 =$
10. $9 \times 9 \times 9 \times 9 \times 9 \times 9 =$

1.
2.
3.
4.
5.
6.
7.
8.
9.
10.
Score

Problem Solving

A number to the third power is equal to -27. What is the number?

Review Exercises

1. $-36 \div 4 =$
2. $-9 - -6 =$
3. $-\frac{1}{3} + -\frac{1}{4} =$
4. $-2.7 + 6.3 =$
5. $-3.12 \div 3 =$
6. $\frac{3}{4} \times -\frac{1}{2} =$

Helpful Hints

$\sqrt{}$ is the symbol for **square root**.
$\sqrt{36}$ is read "the square root of 36."
The answer is the number that when multiplied by itself equals 36.
$\sqrt{36} = 6$, because $6 \times 6 = 36$.
$\sqrt{49} = 7$, because $7 \times 7 = 49$.
$\sqrt{81} = 9$, because $9 \times 9 = 81$.

Find the square roots of the following numbers.

S1. $\sqrt{25} =$
S2. $\sqrt{144} =$
1. $\sqrt{16} =$
2. $\sqrt{121} =$
3. $\sqrt{1} =$
4. $\sqrt{900} =$
5. $\sqrt{100} =$
6. $\sqrt{400} =$
7. $\sqrt{169} =$
8. $\sqrt{9} =$
9. $\sqrt{256} =$
10. $\sqrt{1,600} =$

1.
2.
3.
4.
5.
6.
7.
8.
9.
10.
Score

Problem Solving

The product of -7 and 5 is added to -6.
Find the number.

Review Exercises

1. $6^2 =$
2. $(-2)^3 =$
3. write $6 \times 6 \times 6 \times 6$ as an exponent
4. $\sqrt{64} =$
5. $\sqrt{169} =$
6. $\sqrt{121} =$

Helpful Hints

Use what you have learned about exponents and square roots to solve the following problems.

Examples:

$\frac{\sqrt{64}}{2^2} = \frac{8}{4} = 2$ $\quad$ $\frac{4^2}{2^3} = \frac{16}{8} = 2$

$\frac{3^3}{\sqrt{9}} = \frac{27}{3} = 9$ $\quad$ $\sqrt{36} \times 3^3 = 6 \times 27 = 162$

Solve each of the following.

S1. $\sqrt{16} \times 3^2 =$

S2. $\frac{4^3}{\sqrt{64}} =$

1. $\frac{\sqrt{100}}{\sqrt{25}} =$
2. $\frac{\sqrt{64}}{(2^3)} =$
3. $\frac{5^3}{\sqrt{25}} =$
4. $2^3 \times \sqrt{121} =$
5. $3^2 \times 4^2 =$
6. $\frac{2^3 \times 4^2}{\sqrt{4}} =$
7. $\sqrt{81} \times \sqrt{36} =$
8. $\frac{2^4}{\sqrt{16}} =$
9. $2^2 \times 3^2 \times \sqrt{16} =$
10. $\frac{3^4}{\sqrt{81}} =$

1.
2.
3.
4.
5.
6.
7.
8.
9.
10.
Score

Problem Solving

5 to the second power added to the square root of what number is equal to 34?

Reviewing Exponents and Square Roots

For 1-6, rewrite each as an exponent.

1. $13 \times 13 \times 13 \times 13 =$
2. $2 \times 2 \times 2 \times 2 \times 2 \times 2 \times 2 =$
3. $64 =$
4. $(-2) \times (-2) \times (-2) \times (-2) =$
5. $8 =$
6. $100 =$

For 7-12, find the square root of each number.

7. $\sqrt{16} =$
8. $\sqrt{64} =$
9. $\sqrt{16 + 9} =$
10. $\sqrt{400}$
11. $\sqrt{9}$
12. $\sqrt{4} \times 9 =$

Solve each of the following

13. $\sqrt{36} + 4^2 =$
14. $\dfrac{\sqrt{64}}{\sqrt{4}} =$
15. $6^2 + 7^2 =$
16. $(4^2) \times (5^2) =$
17. $\sqrt{49} \times \sqrt{81} =$
18. $3^2 \times 5^2 \times \sqrt{9} =$
19. $\dfrac{5^3}{5} =$
20. $\dfrac{\sqrt{100} \times \sqrt{25}}{\sqrt{25}} =$

1.
2.
3.
4.
5.
6.
7.
8.
9.
10.
11.
12.
13.
14.
15.
16.
17.
18.
19.
20.

Review Exercises

1. $7^2 =$

2. $\sqrt{36} =$

3. $-9 - -7 =$

4. $16 + -72$

5. $\frac{16 \div -2}{-4 \times -2} =$

6. $7^2 - 5^2 =$

Helpful Hints

It is necessary to follow the correct **order of operations** when simplifying an expression.

1. Evaluate within grouping symbols.
2. Eliminate all exponents.
3. Multiply and divide in order from left to right.
4. Add and subtract in order from left to right.

Examples:

$3^2 (3 + 5) + 3$
$= 3^2 (8) + 3$
$= 9 (8) + 3$
$= 72 + 3$
$= 75$

Sometimes there are no grouping symbols.

$4 + 12 \times 3 - 8 \div 4$
$= 4 + 36 - 2$
$= 40 - 2$
$= 38$

*A number next to a grouping symbol means multiply. $3 (2 + 1) = 3 \times (2 + 1)$

Solve each of the following. Be sure to follow the correct order of operations.

S1. $5 + 9 \times 3 - 4 =$

S2. $8 + 3^2 \times 4 - 6 =$

1. $4 (6 + 2) - 5^2 =$

2. $(14 - 6) + 56 \div 2^3 =$

3. $5^2 + (15 + 3) \div 2 =$

4. $7 \times 4 - 9 \div 3 =$

5. $(3 \times 12) \div (9 \div 3) =$

6. $5^2 + 2^3 - 2 \times 3 =$

7. $12 - 6 \div 3 + 4 =$

8. $3^2 - 2^3 + 6 \div 2 =$

9. $(3 + 8 \div 2) \times (2 \times 6 \div 3) =$

10. $9 + [(4 + 5) \times 3] =$

1.
2.
3.
4.
5.
6.
7.
8.
9.
10.
Score

Problem Solving

A running back gained 12 yards. The next play he lost 18 yards, and on the third play he gained five yards. What was his net gain or net loss?

Review Exercises

Use the following sets to find the answers.

A = {1,5,7,8,9}, B = {2,4,6,8,10}, and C = {1,2,4,5}

1. Find $B \cap C$.
2. Find $A \cap B$.
3. Find $C \cap \varnothing$.
4. Find $A \cup C$.
5. Find $B \cup C$.
6. Find $(A \cap B) \cup C$.

Helpful Hints

*Remember the correct order of operations:
1. Evaluate within grouping symbols.
2. Eliminate all exponents.
3. Multiply and divide in order from left to right.
4. Add and subtract in order from left to right.

Examples:

$14 \div 2 \times 3 + 4^2 - 1$
$= 14 \div 2 \times 3 + 16 - 1$
$= 7 \times 3 + 16 - 1$
$= 21 + 16 - 1$
$= 37 - 1$
$= 36$

$\frac{5(8-3) - 2^2}{3 + 2(3^2 - 7)} = \frac{5(5) - 2^2}{3 + 2(9-7)} = \frac{5(5) - 4}{3 + 2(2)} = \frac{25 - 4}{3 + 4} = \frac{21}{7} = 3$

S1. $\{(2 + 4) \times 3 + 2\} \div 5 =$

S2. $\frac{7^2 - (-5 + 9)}{2(4^2 - 12) - 3} =$

1. $4^3 - 7(2 + 3) =$

2. $\frac{(12 - 3) + 3^2}{-7 + 2(4 + 1)} =$

3. $\frac{4^2 + 12}{5 + 3(2 + 1)} =$

4. $6(-3 + 9) + -4 =$

5. $\frac{10 + (2 + -6)}{4(2^3 - 6) + -2} =$

6. $63 \div 7 - 3 \times 2 + 4 =$

7. $3\{(2 + 7) \div 3 + 7\} \div 5 =$

8. $(12 + -3) + 75 \div 5^2 =$

9. $20 - 3^2 - 5 \times 2 + 6 =$

10. $3^2 + 2^3 + 14 \div 2 =$

1.
2.
3.
4.
5.
6.
7.
8.
9.
10.
Score

Problem Solving

John started the week with \$64. Each day, Monday through Friday, he spent \$7 for lunch. How much money did he have left at the end of the week?

Review Exercises

1. $3^3 =$

2. $5 + 3 \times 7 + 2 =$

3. $-7 - 6 =$

4. Carefully define "set."

5. $\frac{1}{2} \times -2\frac{1}{2} =$

6. $-.91 + .5 =$

Helpful Hints

For any real numbers a, b, and c, the following properties are true:

Property	Rule	Examples:
1. Identity Property of Addition	$0 + a = a$	$0 + 2 = 2$
2. Identity Property of Multiplication	$1 \times a = a$	$1 \times 7 = 7$
3. Inverse Property of Addition	$a + (-a) = 0$	$5 + (-5) = 0$
4. Inverse Property of Multiplication	$a \times \frac{1}{a} = 1 \quad (a \neq 0)$	$6 \times \frac{1}{6} = 1$
5. Associative Property of Addition	$(a + b) + c = a + (b + c)$	$(2 + 3) + 4 = 2 + (3 + 4)$
6. Associative Property of Multiplication	$(a \times b) \times c = a \times (b \times c)$	$(2 \times 3) \times 4 = 2 \times (3 \times 4)$
7. Commutative Property of Addition	$a + b = b + a$	$5 + 6 = 6 + 5$
8. Commutative Property of Multiplication	$a \times b = b \times a$	$4 \times 3 = 3 \times 4$
9. Distributive Property	$a \times (b + c) = a \times b + a \times c$	$5 \times (3 + 2) = 5 \times 3 + 5 \times 2$

Name the property that is illustrated.

S1. $7 + 9 = 9 + 7$

S2. $3 \times (7 + 4) = 3 \times 7 + 3 \times 4$

1. $7 + (-7) = 0$

2. $3 \times (4 \times 5) = (3 \times 4) \times 5$

3. $0 + (-6) = -6$

4. $5 \times \frac{1}{5} = 1$

5. $9 + (6 + 5) = 9 + (5 + 6)$

6. $9 \times 7 = 7 \times 9$

7. $(6 + 5) + 7 = 6 + (5 + 7)$

8. $1 \times \frac{7}{8} = 7$

9. $3 \times 2 + 3 \times 4 = 3 \times (2 + 4)$

10. $16 + (-16) = 0$

1.
2.
3.
4.
5.
6.
7.
8.
9.
10.

Problem Solving

Five times a certain number is equal to 95. Find the number.

Score

Review Exercises

1. $3 + 6 \times 2 - 4 =$
2. $3\,(5 + 2) - 2^2 =$
3. $3 \times 4 - 6 \div 3 =$
4. $2^2 + 3^3 - 2 \times 4 =$
5. $(4 + 4 \div 2) \times (2 \times 10 \div 2) =$
6. $3\,[(4 + 3) \times 2] =$

Helpful Hints

Use what you have learned to solve the following problems.

Example: Use the indicated property to complete the statement with the correct answer.

Inverse Property of Addition:	$27 + (\ \ ) = 0$	answer: -27
Distributive Property:	$4\,(5 + 7) =$	answer: $4 \times 5 + 4 \times 7$

Use the indicated property to complete the statement with the correct answer.

S1. Associative Property of Addition: $(3 + 7) + 9 =$

S2. Commutative Property of Multiplication: $7 \times 15 =$

1. Inverse Property of Multiplication: $9 \times (\ \ ) = 1$
2. Distributive Property: $3 \times (6 + 2) =$
3. Commutative Property of Addition: $9 + 12 =$
4. Associative Property of Multiplication: $3 \times (9 \times 5) =$
5. Distributive Property: $3 \times 5 + 3 \times 7 =$
6. Inverse Property of Addition: $9 + (\ \ ) = 0$
7. Identity Property of Multiplication: $7 \times (\ \ ) = 7$
8. Inverse Property of Multiplication: $\frac{1}{5}(\ \ ) = 1$
9. Associative Property of Addition: $3 + (5 + 6) =$
10. Distributive Property: $6 \times (4 + (-2)) =$

1.
2.
3.
4.
5.
6.
7.
8.
9.
10.
Score

Problem Solving

Mr. Andrews rents a car for one day. He pays $30 per day for the rental plus $.30 per mile he drives. How much will the total price of the rental car be if he drives 40 miles?

Review Exercises

1. $\sqrt{100}$

2. $5^3 =$

3. $\frac{\sqrt{36} \times \sqrt{49}}{2} =$

4. $-3 \times -5 \times -6 =$

5. $-2\frac{1}{2} \div -\frac{1}{2} =$

6. $-\frac{1}{3} + -\frac{1}{2} =$

Helpful Hints

Scientific notation is used to express very large and very small numbers. A number in scientific notation is expressed as the product of two factors. The first factor is a number between 1 and 10 and the second factor is a power of 10 as in the examples $\mathbf{2.346 \times 10^5}$ and $\mathbf{3.976 \times 10^{-7}}$.

Example for a large number: Change 157,000,000,000 to scientific notation. Move the decimal between the 1 and the 5. Since the decimal has moved 11 places to the **left**, the answer is 1.57×10^{11}.

Example for a small number: Change .0000000468 to scientific notation. Move the decimal between the 4 and the 6. Since the decimal has moved eight places to the **right**, the answer is 4.68×10^{-8}.

Change the following numbers to scientific notation.

S1. 2,360,000,000

S2. .000000149

1. 653,000,000,000

2. 159,700

3. 106,000,000

4. .000007216

5. 1,096,000,000

6. .001963

7. .00000000016

8. .0000000008

9. 7,000,000,000,000

10. .0000001287

1.
2.
3.
4.
5.
6.
7.
8.
9.
10.
Score

Problem Solving

Light travels at 186,000 miles per second. Write this speed in scientific notation.

Review Exercises

1. Change 123,000 to scientific notation.

2. Change .000321 to scientific notation.

3. Which property of numbers is illustrated? $3 \times 5 + 3 \times 7 = 3 \times (5 + 7)$

4. $-9 - 8 =$

5. $\dfrac{36 \div -3}{-16 \div -4} =$

6. $2^3 + 3^3 =$

Helpful Hints

It is easy to change numbers in scientific notation to conventional numbers.

Examples:

Change 3.458×10^8 to a conventional number.
Move the decimal eight places to the right. The answer is 345,800,000.

Change 4.5677×10^{-7} to a conventional number.
Move the decimal seven spaces to the left. The answer is .00000045677.

Change each number in scientific notation to a conventional number.

S1. 7.032×10^6

S2. 5.6×10^{-5}

1. 2.3×10^5

2. 9.13×10^{-8}

3. 1.2362×10^{-5}

4. 5.17×10^{11}

5. 1.127×10^3

6. 3.012×10^{-3}

7. 6.67×10^6

8. 2.1×10^4

9. 7×10^{-8}

10. 8×10^6

1.
2.
3.
4.
5.
6.
7.
8.
9.
10.
Score

Problem Solving

The distance to the sun is approximately 9.3×10^7 miles. Change this distance to a conventional number.

Review Exercises

1. Change 123,000 to scientific notation.
2. Change .0000056 to scientific notation.
3. Change 2.76×10^6 to a conventional number.
4. Change 3.75×10^{-5} to to a conventional number.
5. List two equivalent sets.
6. List two disjoint sets.

Helpful Hints

A **ratio** compares two numbers or groups of objects.

Example: ○ ○ ○ / □ □ □ □ — For every three circles there are four squares.

The ratio can be written in the following ways: 3 to 4, 3 : 4, and $\frac{3}{4}$. Each of these is read as "three to four."

*Ratios are often written in fraction form. The first number mentioned is the numerator. Ratios that are expressed as fractions can be reduced to lowest terms.

Write each of the following ratios as a fraction reduced to lowest terms.

S1. 5 nickels to 3 dimes

S2. 18 horses to 4 cows

1. 7 to 2
2. 6 children to 5 adults
3. 30 books to 25 pencils
4. 15 bats to 3 balls
5. 24 to 20
6. 16 to 12
7. 7 dimes to 3 pennies
8. 6 chairs to 4 desks
9. 4 cats to 8 dogs
10. 9 : 3

1.
2.
3.
4.
5.
6.
7.
8.
9.
10.
Score

Problem Solving

A team won 24 games and lost 10. Write the ratio of games won to games lost as a fraction reduced to lowest terms.

Review Exercises

1. Write .00027 in scientific notation.
2. Write 2,916,000 in scientific notation.
3. Write 7.21×10^5 as a conventional number.
4. Write $6.23 \times 10\text{-}5$ as a conventional number.
5. $(-3) \times 2 \times (-5) =$
6. $-.264 \div .2 =$

Helpful Hints

Two equal ratios can be written as a **proportion**.

Example: $\frac{4}{6} = \frac{2}{3}$ In a proportion, the cross products are equal.

Examples: Is $\frac{3}{4} = \frac{5}{6}$ a proportion? To find out, cross multiply.

$3 \times 6 = 18$, $4 \times 5 = 20$, $18 \neq 20$. It is not a proportion.

Is $\frac{6}{9} = \frac{8}{12}$ a proportion? To find out, cross multiply.

$6 \times 12 = 72$, $9 \times 8 = 72$, $72 = 72$. It is a proportion.

Cross multiply to determine whether each of the following is a proportion.

S1. $\frac{2}{5} = \frac{6}{15}$

S2. $\frac{18}{24} = \frac{4}{5}$

1. $\frac{7}{14} = \frac{3}{6}$

2. $\frac{5}{3} = \frac{14}{9}$

3. $\frac{18}{2} = \frac{27}{3}$

4. $\frac{4}{5} = \frac{12}{15}$

5. $\frac{15}{20} = \frac{6}{8}$

6. $\frac{5}{2} = \frac{11}{4}$

7. $\frac{2}{5} = \frac{12}{30}$

8. $\frac{3}{1.3} = \frac{9}{3.5}$

9. $\frac{1/4}{4} = \frac{1/2}{8}$

10. $\frac{5}{8} = \frac{6}{7}$

1.
2.
3.
4.
5.
6.
7.
8.
9.
10.
Score

Problem Solving

A whole number to the power of three, added to five, equals 13. Find the whole number.

Review Exercises

1. Write 25 to 15 as a fraction in lowest terms.

2. Is $\frac{4}{5} = \frac{8}{10}$ a proportion? Why?

3. Is $\frac{2}{5} = \frac{5}{7}$ a proportion? Why?

4. $\sqrt{49} + 3^2 =$

5. $4^3 - 2^4 =$

6. $-225 + 500 =$

Helpful Hints

It is easy to find the missing number in a proportion.

Examples: Solve each proportion.

$\frac{4}{n} = \frac{2}{3}$

First, cross multiply: $2 \times n = 4 \times 3$
$2 \times n = 12$

Next, divide 12 by 2: $2\overline{)12}$ = 6 ; $n = 6$

$\frac{4}{5} = \frac{y}{7}$

First, cross multiply: $5 \times y = 4 \times 7$
$5 \times n = 28$

Next, divide 28 by 5: $5\overline{)28}$ = $5\frac{3}{5}$; $y = 5\frac{3}{5}$

Find the missing number in each proportion.

S1. $\frac{3}{15} = \frac{n}{5}$

S2. $\frac{4}{7} = \frac{x}{28}$

1. $\frac{n}{4} = \frac{12}{16}$

2. $\frac{x}{40} = \frac{5}{100}$

3. $\frac{1}{3} = \frac{14}{y}$

4. $\frac{n}{4} = \frac{8}{5}$

5. $\frac{15}{20} = \frac{n}{8}$

6. $\frac{7}{n} = \frac{3}{9}$

7. $\frac{27}{3} = \frac{n}{2}$

8. $\frac{n}{2} = \frac{7}{5}$

9. $\frac{n}{1.4} = \frac{6}{7}$

10. $\frac{7}{n} = \frac{21}{7}$

1.
2.
3.
4.
5.
6.
7.
8.
9.
10.
Score

Problem Solving

The temperature at midnight is -12°. By 6:00 a.m., the temperature has dropped another 20°. What is the temperature at 6:00 a.m.?

Review Exercises

1. Is $\frac{3}{4} = \frac{9}{12}$ a proportion? Why?

2. Solve the proportion: $\frac{n}{12} = \frac{5}{2}$

3. Solve the proportion: $\frac{5}{6} = \frac{10}{n}$

4. Write 234,000,000 in scientific notation.

5. Write .00235 in scientific notation.

6. Write 7.2×10^5 as as a conventional number.

Helpful Hints

Ratios and **proportions** can be used to solve problems.

Example: A car can travel 384 miles in six hours. How far can the car travel in eight hours?

First set up a proportion. $\frac{384 \text{ miles}}{6 \text{ hours}} = \frac{n \text{ miles}}{8 \text{ hours}}$

Next, cross multiply. $6 \times n = 8 \times 384$
$6 \times n = 3{,}072$

Next, divide by six. $3072 \div 6 = 512$ — $n = 512$

The car can travel 512 miles in eight hours.

Use a proportion to solve each problem.

S1. A car can travel 85 miles on five gallons of gas. How far can the car travel on 12 gallons of gas?

S2. If two pounds of beef cost $4.80, how much will five pounds cost?

1. A car can travel 100 miles on five gallons of gas. How many gallons will be needed to travel 40 miles?

2. Two pounds of chicken cost $7. How much will five pounds cost?

3. In a class, the ratio of boys to girls is four to three. If there are 20 boys in the class, how many girls are there?

4. A runner takes three hours to go 24 miles. At this rate, how far could he run in five hours?

5. Seven pounds of nuts cost $5. How many pounds of nuts can you buy with $2?

1.
2.
3.
4.
5.
Score

Problem Solving

At 6:00 a.m. the temperature was -16°. By noon the temperature had risen 28°. What was the temperature at noon?

Reviewing Ratios and Proportions

For 1-3, write each ratio as a fraction reduced to lowest terms.

1. 12 to 4
2. 24 to 10
3. 16 to 6

For 4-6, determine whether each is a proportion and why.

4. $\frac{12}{15} = \frac{24}{30}$
5. $\frac{7}{8} = \frac{8}{9}$
6. $\frac{5}{3} = \frac{15}{9}$

For 7-15, solve each proportion.

7. $\frac{12}{15} = \frac{n}{5}$
8. $\frac{1}{3} = \frac{11}{n}$
9. $\frac{1}{20} = \frac{n}{100}$
10. $\frac{5}{7} = \frac{25}{n}$
11. $\frac{3}{4} = \frac{n}{6}$
12. $\frac{15}{20} = \frac{n}{12}$
13. $\frac{10}{100} = \frac{2}{n}$
14. $\frac{x}{5} = \frac{9}{15}$
15. $\frac{3}{16} = \frac{n}{48}$

For 16-20, use a proportion to solve each problem.

16. If four pounds of pork cost $4.80, how much will seven pounds cost?
17. In a class the ratio of girls to boys is two to three. If there are 20 girls, how many boys are in the class?
18. A cyclist can travel 42 miles in three hours. How far can he travel in five hours?
19. A car can travel 120 miles on five gallons of gas. How many gallons will be needed to travel 48 miles?
20. If six pounds of nuts cost $18, how many pounds of nuts can you buy with $12?

1.
2.
3.
4.
5.
6.
7.
8.
9.
10.
11.
12.
13.
14.
15.
16.
17.
18.
19.
20.

Review Exercises

1. Solve the proportion: $\frac{7}{6} = \frac{n}{18}$

2. Solve the proportion: $\frac{n}{3} = \frac{6}{5}$

3. $3 \times 2 + 6 \div 2 =$

4. $4^2 + (5 \times 2) \div 5 =$

5. $4^2 + 2^2 + 12 \div 2 =$

6. $5(-2 + -6) + 7 =$

Helpful Hints

Percent means "**per hundred**" or "**hundredths.**"

Percents can be expressed as decimals and as fractions.
The fraction form may sometimes be reduced to its lowest terms.

Examples: $25\% = .25 = \frac{25}{100} = \frac{1}{4}$ $\quad 8\% = .08 = \frac{8}{100} = \frac{2}{25}$

Change each percent to a decimal and to a fraction reduced to its lowest terms.

S1. 20% = _____

S2. 9% = _____

1. 16% = _____

2. 6% = _____

3. 75% = _____

4. 40% = _____

5. 1% = _____

6. 45% = _____

7. 12% = _____

8. 5% = _____

9. 50% = _____

10. 13% = _____

1.
2.
3.
4.
5.
6.
7.
8.
9.
10.
Score

Problem Solving

95% of the students enrolled in a school are present.
What fraction are present? (Reduce to lowest terms.)

Review Exercises

1. Change 80% to a decimal.

2. Change 7% to a decimal.

3. Change 25% to a fraction reduced to lowest terms.

4. $156 \times .7$

5. $400 \times .32$

6. $300 \times .06$

Helpful Hints

To find the **percent of a number**, you may use either fractions or decimals. Use what is the most convenient.

Example: Find 25% of 60.
.25 × 60

$$\begin{array}{r} 60 \\ \times .25 \\ \hline 300 \\ 120 \\ \hline 15.00 \end{array}$$

OR

$$\frac{25}{100} = \frac{1}{4}$$

$$\frac{1}{\cancel{4}_1} \times \frac{\cancel{60}^{15}}{1} = \frac{15}{1} = 15$$

S1. Find 70% of 25.

S2. Find 50% of 300.

1. Find 6% of 72.

2. Find 60% of 85.

3. Find 25% of 60.

4. Find 45% of 250.

5. Find 10% of 320.

6. Find 40% of 200.

7. Find 4% of 250.

8. Find 90% of 240.

9. Find 75% of 150.

10. Find 2% of 660.

1.
2.
3.
4.
5.
6.
7.
8.
9.
10.
Score

Problem Solving

Arlene took a test with 40 questions. If she got a score of 85% correct, how many problems did she get correct?

Review Exercises

1. Find 15% of 310.
2. Find 20% of 120.
3. $8\overline{)6}$
4. Change .7 to a percent.
5. Find .9 of 150.
6. $.05\overline{)30}$

Helpful Hints

When finding the **percent**, first write a fraction, change the fraction to a decimal, and then change the decimal to a percent.

Examples: 4 is what percent of 16?

$\frac{4}{16} = \frac{1}{4}$

$4\overline{)1.00}$ = .25 = 25%
- 8
20
- 20
0

5 is what percent of 25?

$\frac{5}{25} = \frac{1}{5}$

$5\overline{)1.00}$ = .20 = 20%
- 1 0
00

S1. 3 is what percent of 12?

S2. 15 is what percent of 20?

1. 7 is what percent of 28?
2. 20 is what percent of 25?
3. 40 = what percent of 80?
4. 18 is what percent of 20?
5. 12 is what percent of 20?
6. 9 is what percent of 12?
7. 15 = what percent of 20?
8. 24 is what percent of 32?
9. 400 is what percent of 500?
10. 19 is what percent of 20?

1.
2.
3.
4.
5.
6.
7.
8.
9.
10.
Score

Problem Solving

A rancher had 800 cows. He sold 600 of them.
What percent of the cows did he sell?

Review Exercises

1. Find 4% of 80.
2. Find 40% of 80.
3. Twelve is what percent of 16?
4. 45 is what percent of 50?
5. 52 - 1.96 =
6. $.06\overline{)12}$

Helpful Hints

To find the **whole** when the **part** and the **percent** are known, simply change the equal sign (" = ") to the division sign (" ÷ ").

Examples: 6 = 25% of what number?
6 ÷ 25% (Change = to ÷.)
6 ÷ .25 (Change % to decimal.)
$.25\overline{)6.00}$ = 24.

Twelve is 80% of what?
12 ÷ 80% (Change = to ÷.)
12 ÷ .8 (Change % to decimal.)
$.8\overline{)12.0}$ = 15.

* Be careful to move decimal points properly.

S1. 5 = 25% of what?

S2. Six is 20 % of what?

1. 12 = 25% of what?
2. 32 = 40 % of what?
3. Five is 20% of what?
4. 3 = 75% of what?
5. Twelve is 80% of what?
6. 8 = 40% of what?
7. 15 is 25% of what?
8. Fifteen is 20% of what?
9. 9 is 20% of what?
10. 25 is 20% of what?

1.
2.
3.
4.
5.
6.
7.
8.
9.
10.
Score

Problem Solving

There are 15 girls in a class. If this is 60% of the class, how many students are there in the class?

Review Exercises

1. Change $\frac{72}{100,000}$ to a decimal.
2. Change 2.0019 to a mixed numeral.
3. Change $\frac{9}{15}$ to a percent.
4. $\frac{3}{5} \times 25 =$
5. $8\overline{).168}$
6. $.03\overline{)2.4}$

Helpful Hints

Use what you have learned to solve the following word problems. **Examples:**

A man earns $300 and spends 40% of it. How much does he spend?

Find 40% of 300.

$$\begin{array}{r} 300 \\ \times .4 \\ \hline 120 \end{array}$$

He spends $120.

In a class of 25 students, 15 are girls. What % are girls?

15 = what % of 25

$$\frac{15}{25} = \frac{3}{5}$$

$$5\overline{)3.00} = .60$$

60% are girls.

Five students got A's on a test. This is 20% of the class. How many are in the class?

5 = 20% of what?

5 ÷ .2

$$.2\overline{)5.0} = 25.$$

25 are in the class.

S1. On a test with 25 questions, Al got 80% correct. How many questions did he get correct?

S2. A player took 12 shots and made 9. What percent did he make?

1. A girl spent $5. This was 20% of her earnings. How much were her earnings?
2. Buying a $8,000 car requires a 20% down payment. How much is the down payment?
3. 3 = 10% of what?
4. A team played 20 games and won 18. What % did they win?
5. A farmer sold 50 cows. If this was 20% of his herd, how many cows were in his herd?
6. 20 = 80% of what?
7. Paul wants a bike that costs $400. If he has saved 60% of this amount, how much has he saved?
8. There are 400 students in a school. Fifty-five percent are girls. How many boys are there?
9. 12 is what % of 60?
10. Kelly earned 300 dollars and put 70% of it into the bank. How much did she put into the bank?

1.
2.
3.
4.
5.
6.
7.
8.
9.
10.
Score

Problem Solving

Nacho's monthly income is $4,800. What is his annual income? (Hint: How many months are there in a year?)

Review Exercises

1. $7.68 + 19.7 + 5.364 =$

2. $\begin{array}{r} 7.123 \\ -\ 4.765 \\ \hline \end{array}$

3. $\begin{array}{r} 3.14 \\ \times\ 7 \\ \hline \end{array}$

4. $\begin{array}{r} .208 \\ \times\ .06 \\ \hline \end{array}$

5. $3\overline{)1.44}$

6. $.15\overline{)1.215}$

Helpful Hints

Use what you have learned to solve the following problems.

*Refer to the examples on the previous page if necessary.

S1. Find 20% of 150.

S2. 6 is 20% of what?

1. 8 is what % of 40?

2. Change $\frac{18}{20}$ to a percent.

3. A school has 600 students. If 5% are absent, how many students are absent?

4. A quarterback threw 24 passes and 75% of them were caught. How many were caught?

5. Riley has 250 marbles in his collection. If 50 of them are red, what percent of them are red?

6. A team played 60 games and won 45 of them. What % did they win?

7. There are 50 sixth graders in a school. This is 20% of the school. How many students are in the school?

8. A coat is on sale for $20. This is 80% of the regular price. What is the regular price?

9. Steve has finished $\frac{3}{5}$ of his test. What percent of the test has he finished?

10. Alex wants to buy a computer priced at $640. If sales tax is 8%, what is the total cost of the computer?

1.
2.
3.
4.
5.
6.
7.
8.
9.
10.
Score

Problem Solving

Ann took five tests and scored a total of 485 points. What was her average score?

Reviewing Percents

Change numbers 1 - 5 to percents.

1. $\frac{13}{100} =$ 2. $\frac{3}{100} =$ 3. $\frac{7}{10} =$ 4. .19 = 5. .6 =

Change numbers 6 - 8 to a decimal and a fraction expressed in lowest terms.

6. 8% = ._____ = _____ 7. 18% = ._____ = _____ 8. 80% = ._____ = _____

Solve the following problems. Label the word problem answers.

9. Find 3% of 74.

10. Find 40% of 320.

11. 20 is what percent of 25?

12. 15 is what percent of 20?

13. 3 = 25% of what?

14. 15 = 20% of what?

15. Change $\frac{16}{20}$ to a %.

16. Change $\frac{3}{5}$ to a percent.

17. 640 students attend Lincoln School. If 40% of the students are girls, how many girls attend Lincoln School?

18. A team played 40 games. If they won 65% of them, how many games did the team win?

19. A pitcher threw 40 pitches. If 30 were strikes, what percent were strikes?

20. Thirty students attended an assembly. This was 20% of the seventh grade. How many students are there in the seventh grade?

Answers
1.
2.
3.
4.
5.
6.
7.
8.
9.
10.
11.
12.
13.
14.
15.
16.
17.
18.
19.
20.

Review Exercises

1. Change .7 to a percent.
2. Change $\frac{4}{5}$ to a percent.
3. Change .12 to a fraction reduced to lowest terms.
4. Find 6% of 200.
5. Three is what percent of 12?
6. 5 = 20% of what?

Helpful Hints

A **factor** of a whole number is a whole number that divides into it evenly, without a remainder.

Examples: Find all factors of 20.

$1 \times 20 = 20$ $2 \times 10 = 20$ $4 \times 5 = 20$

All the factors of 20 are: 1, 20, 2, 20, 4, 5

Find all factors of 84.

$1 \times 84 = 84$ $2 \times 42 = 84$ $3 \times 28 = 84$

$4 \times 21 = 84$ $6 \times 14 = 84$ $7 \times 12 = 84$

All the factors of 84 are: 1, 84, 2, 42, 3, 28, 4, 21, 6, 14, 7, 12

Find all the factors of each number.

S1. 30
S2. 36
1. 100
2. 42
3. 70
4. 81
5. 50
6. 40
7. 75
8. 90
9. 20
10. 50

1.
2.
3.
4.
5.
6.
7.
8.
9.
10.
Score

Problem Solving

A test contained 60 questions. If a student's score was 90%, how many questions did he get correct?

Review Exercises

1. $-9 - 6 + -3 =$

2. $-3 \times -2 \bullet 4 =$

3. $\sqrt{121} + \sqrt{81}$

4. Solve the proportion. $\frac{3}{4} = \frac{n}{10}$

5. 3 = 20% of what?

6. Two is what % of eight?

Helpful Hints

The **greatest common factor** is the largest factor that two or more numbers have in common.

Example: Find the greatest common factor of 12 and 16.

Find the factors of each number: 12: 1, 2, 3, (4), 6, 12
16: 1, 2, (4), 8, 16 greatest common factor = (4)

* "Greatest common factor" is abbreviated as GCF.

Find the greatest common factor of each pair of numbers.

S1. 8 and 10

S2. 12 and 20

1. 6 and 8

2. 12 and 15

3. 42 and 56

4. 64 and 80

5. 100 and 120

6. 90 and 70

7. 45 and 25

8. 60 and 72

9. 48 and 36

10. 20 and 40

1.
2.
3.
4.
5.
6.
7.
8.
9.
10.
Score

Problem Solving

Light travels approximately 5.879×10^{12} miles in one year. Write the distance travelled as a conventional number.

Review Exercises

1. Write .0000012 in scientific notation.

2. Write 496,000,000 in scientific notation.

3. Write 1.32×10^7 as a conventional number.

4. Write 4.64 × 10-6 as a conventional number.

5. Find all the factors of 60.

6. Find the GCF (greatest common factor) of 36 and 40.

Helpful Hints

A **multiple** of a number is the product of that number and any whole number.

The multiples of a number can be found by multiplying it by 0, 1, 2, 3, 4, and so on.

Example: Find the first six multiples of 3.
3: 0, 3, 6, 9, 12, 15

These are found by multiplying 3 by 0, 1, 2, 3, 4, and 5.

Complete the list of multiples for each number.

S1. 2: 0, 2, ☐, ☐, ☐, ☐

S2. 6: ☐, 6, ☐, ☐, 24, ☐

1. 5: 0, 5, ☐, ☐, ☐, ☐

2. 3: ☐, 3, ☐, 9, ☐, ☐

3. 10: ☐, 10, 20, ☐, ☐, ☐

4. 4: ☐, ☐, ☐, 12, 16, 20

5. 11: 0, 11, ☐, 33, ☐, 55

6. 8: 0, 8, 16, ☐, ☐, ☐

7. 20: 0, 20, 40, ☐, ☐, ☐

8. 7: 0, 7, ☐, 21, ☐, ☐

9. 30: 0, 30, 60, ☐, ☐, ☐

10. 9: 0, 9, 18, ☐, 36, ☐

1.
2.
3.
4.
5.
6.
7.
8.
9.
10.
Score

Problem Solving

A pitcher threw 30 pitches that were strikes.
This was 25% of all the pitches thrown.
How many pitches were thrown by the pitcher?

Review Exercises

1. List all the factors of 30.
2. Find the GCF of 32 and 60.
3. List the first six multiples of eight.
4. Find 6% of 50.
5. 3 is what % of 12?
6. 7 = 20% of what?

Helpful Hints

The **least common multiple** of two or more whole numbers is the smallest whole number, other than zero, that they all divide into evenly.

Examples: The least common multiple of:
2 and 3 is 6   4 and 6 is 12   3 and 9 is 9

* Least common multiple is abbreviated as LCM.

Find the least common multiple of each pair of numbers.

S1. 3 and 4

S2. 6 and 8

1. 3 and 5
2. 6 and 10
3. 12 and 20
4. 10 and 15
5. 12 and 18
6. 15 and 60
7. 16 and 12
8. 8 and 20
9. 9 and 12
10. 12 and 30

1.
2.
3.
4.
5.
6.
7.
8.
9.
10.
Score

Problem Solving

A CD costs $12. If the sales tax is 8%, what is the total cost of the CD?

Reviewing Number Theory

For 1-6, find all factors for each number.

1. 24
2. 16
3. 32
4. 28
5. 70
6. 25

For 7-12, find the greatest common factor for each pair of numbers.

7. 12 and 8
8. 48 and 60
9. 120 and 100
10. 45 and 50
11. 35 and 28
12. 90 and 72

For 13 - 15, complete the list of multiples of each number.

13. 3: 0, 3, 6, ☐, ☐, ☐
14. 9: 0, ☐, 18, ☐, ☐, ☐,
15. 15: 0, ☐, ☐, ☐, ☐, 75

For 16-20, find the least common multiple of each pair of numbers.

16. 4 and 6
17. 12 and 15
18. 20 and 15
19. 12 and 4
20. 8 and 6

1.
2.
3.
4.
5.
6.
7.
8.
9.
10.
11.
12.
13.
14.
15.
16.
17.
18.
19.
20.

Review Exercises

1. Find the GCF of 40 and 56.
2. Find the LCM of 4 and 6
3. List all factors of 28.
4. List the first six multiples of 12.
5. Is $\frac{6}{8} = \frac{3}{4}$ a proportion? Why?
6. Solve the proportion. $\frac{6}{8} = \frac{n}{12}$

Helpful Hints

Numbers can be assigned to a point on a **number line**. **Positive numbers** are to the right of zero. **Negative numbers** are to the left of zero.

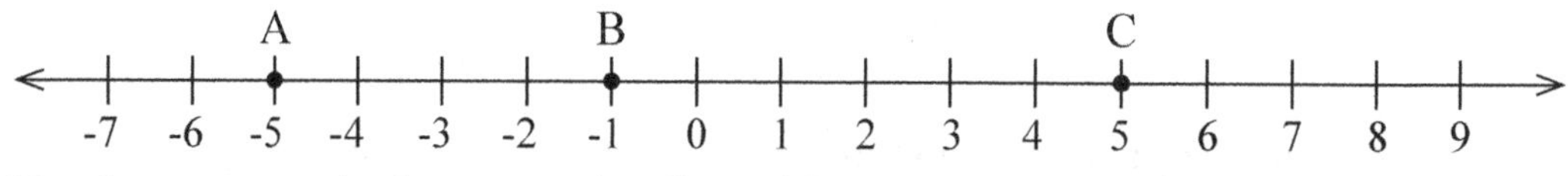

Numbers are graphed on a number line with a point.

Examples: A is the graph of -5. B is the graph of -1. C is the graph of 5.

A has a coordinate of -5. B has a coordinate of -1. C has a coordinate of 5.

Use the number line to state the coordinates of the given points.

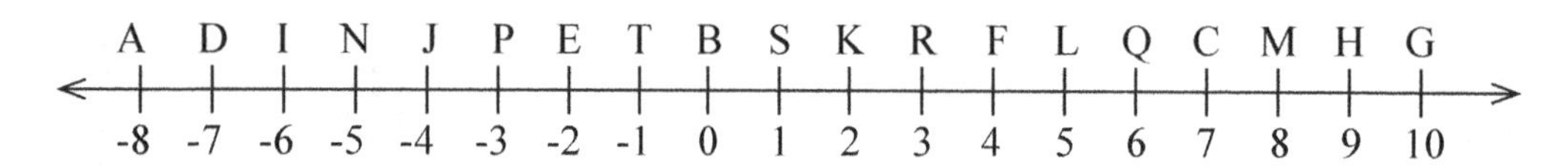

S1. B

S2. D, E, and G

1. L and H
2. R and F
3. K, F, and C
4. N and A
5. G, H, I, and Q
6. H, D, and S
7. A, M, B, and P
8. B, C, and M
9. I, F, and P
10. L, P, H, and A

1.
2.
3.
4.
5.
6.
7.
8.
9.
10.
Score

Problem Solving

At midnight the temperature was -4°.
By 6:00 a.m. the temperature had risen 12°.
What was the temperature at 6:00 a.m.?

Review Exercises

Use A = {2,4,6,8,10}, B = {1,3,4,5,6,8,10}, and C = {4,5,6,8,9,10} to answer the following questions.

1. $A \cap B =$ 2. $B \cup C =$ 3. $A \cup C =$ 4. $B \cap C =$

5. Are B and C equivalent sets? Why?

6. Are A and C disjoint sets? Why?

Helpful Hints

Equations can be solved and graphed on a **number line**.

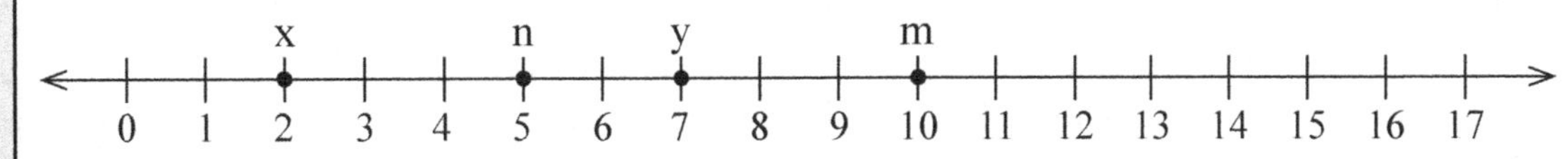

Examples:

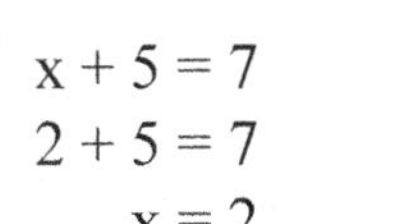

$x + 5 = 7$
$2 + 5 = 7$
$x = 2$

$n - 3 = 2$
$5 - 3 = 2$
$x = 5$

$3y = 21$
$3 \times 7 = 21$
$y = 7$

$\frac{m}{2} = 5$
$\frac{10}{2} = 5$
$m = 10$

Solve each equation and graph each solution on the number line.
Also place each solution in the answer column.

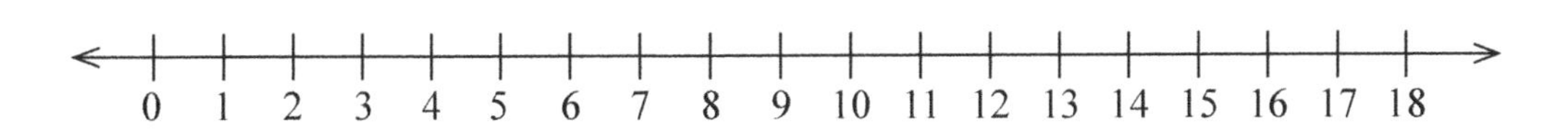

S1. $x + 2 = 3$ S2. $y - 2 = 5$ 1. $c + 4 = 7$

2. $5 - e = 0$ 3. $3d = 15$ 4. $\frac{f}{3} = 6$

5. $n \times 2 = 8$ 6. $3 + j = 14$ 7. $3 + k = 11$

8. $4m = 28$ 9. $6 = n + 2$ 10. $\frac{r}{2} = 6$

1.
2.
3.
4.
5.
6.
7.
8.
9.
10.
Score

Problem Solving

A car can travel 320 miles in five hours.
At this rate, how far can it travel in eight hours?

Review Exercises

1. $-2 + 9 =$
2. $-7 - 15 =$
3. $-7 - -15 =$
4. $6 \times -7 =$
5. $-45 \div -9 =$
6. $\frac{-24 \div -2}{18 \div -3} =$

Helpful Hints

Ordered pairs can be graphed on a **coordinate system**.

The first number of an ordered pair shows how to move across. It is called the **x-coordinate**.

The second number of an ordered pair shows how to move up and down. It is called the **y-coordinate**.

Examples: To locate B, move across to the right to 3 and up 4. The ordered pair is (3,4).

To locate C, move across to the left to -5 and up 2. The ordered pair is (-5,2).

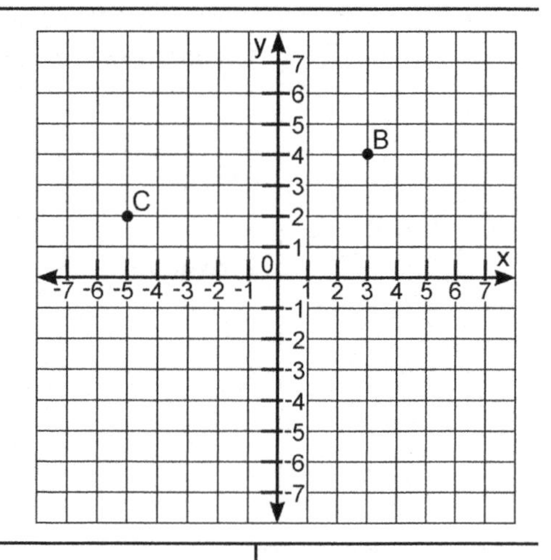

Use the coordinate system to find the point associated with each ordered pair.

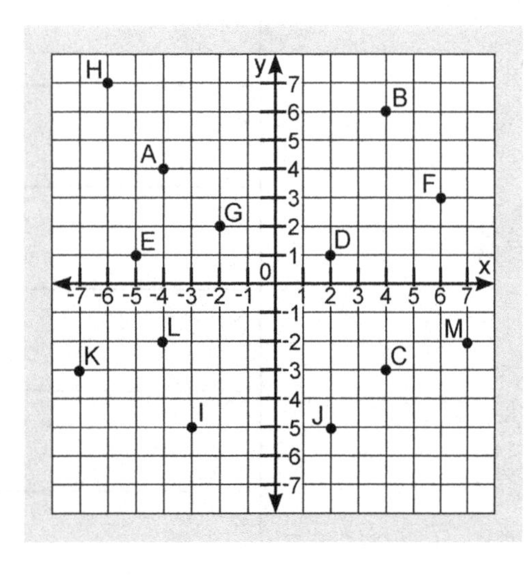

S1. D

S2. L

1. F
2. J
3. K
4. E
5. B
6. C
7. I
8. G
9. D
10. H

1.

2.

3.

4.

5.

6.

7.

8.

9.

10.

Score

Problem Solving

A shirt that regularly sells for $30 is on sale for 20% off. How much is the sale price?

Review Exercises

1. $\frac{1}{3} + -\frac{4}{5} =$
2. $-.29 + -.39 =$
3. $-\frac{1}{8} - (-\frac{1}{2}) =$
4. $-\frac{2}{3} \times -1\frac{1}{2} =$
5. $2\frac{1}{2} \div -\frac{1}{2} =$
6. $-5 \div -2\frac{1}{2} =$

Helpful Hints

A **point** can be found by matching it with an ordered pair.

Examples: (-5, 3) is found by moving across to the left to -5, and up 3. This is represented by point B. -5 is the **x-coordinate** and 3 is the **y-coordinate**.

(6, 3) is found by moving across to the right to 6, and up 3. This is represented by point C. 6 is the **x-coordinate** and 3 is the **y-coordinate**.

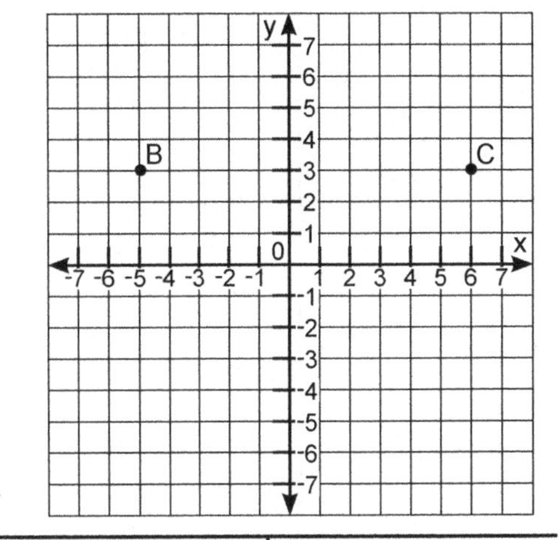

Use the coordinate system to find the point associated with each ordered pair.

S1. (6, 2)   S2. (-5, 5)

1. (3, 5)
2. (7, -6)
3. (-6, -4)
4. (0, 3)
5. (-2, -4)
6. (-2, 2)
7. (-6, 2)
8. (4, 2)
9. (-4, -7)
10. (4, -3)

1.
2.
3.
4.
5.
6.
7.
8.
9.
10.

Score

Problem Solving

In a class of 40 students, 38 were present.
What percent of the class was present?

Review Exercises

1. $2^5 =$
2. $\sqrt{36} + 4^2 =$
3. $\dfrac{4^2 + 3^2}{\sqrt{25}} =$
4. $2^3 \times 3^2 =$
5. Write .00017 in scientific notation.
6. Write 213,000 in scientific notation.

Helpful Hints

The **slope** of a line refers to how steep the line is. It is the ratio of **rise to run**.

$$\text{slope} = \frac{y_2 - y_1}{x_2 - x_1}$$

Example:
What is the slope of the line passing through the ordered pairs (1, 5) and (6, 9)?

$$\text{slope} = \frac{y_2 - y_1}{x_2 - x_1} \qquad \overset{x_1\ y_1}{(1, 5)}, \overset{x_2\ y_2}{(6, 9)}$$

$$= \frac{9 - 5}{6 - 1}$$

$$= \frac{4}{5}$$

The slope is $\frac{4}{5}$

The run is 5 and the rise is 4.

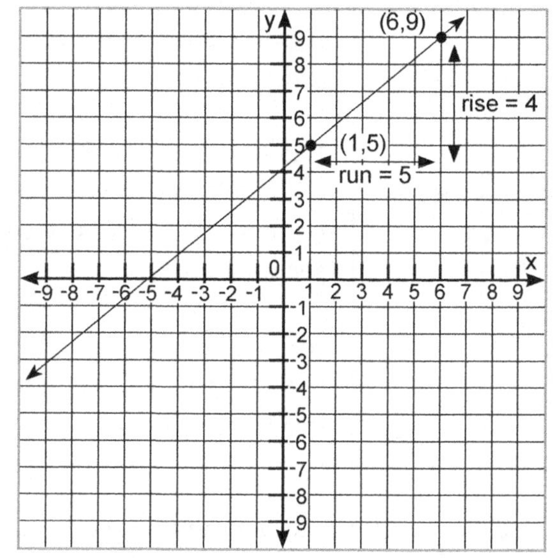

Find the slope of each line that passes through the given point.

S1. (2, 3), (5, 4)

S2. (3, -2), (5, 1)

1. (4, 3), (2, 6)
2. (4, 1), (7, 2)
3. (-2, 1), (-3, 3)
4. (-2, -2), (6, 3)
5. (4, 5), (6, 6)
6. (1, 2), (3, 9)
7. (1, -1), (6, 5)
8. (3, 2), (8, 6)
9. (2, -1), (4, 2)
10. (9, 2), (7, 5)

1.
2.
3.
4.
5.
6.
7.
8.
9.
10.
Score

Problem Solving

In a school the ratio of boys to girls is five to four. If there are 400 boys, how many girls are there in the school?

Reviewing Number Lines and Coordinate Systems

Use the number line to state the coordinates of the given points.

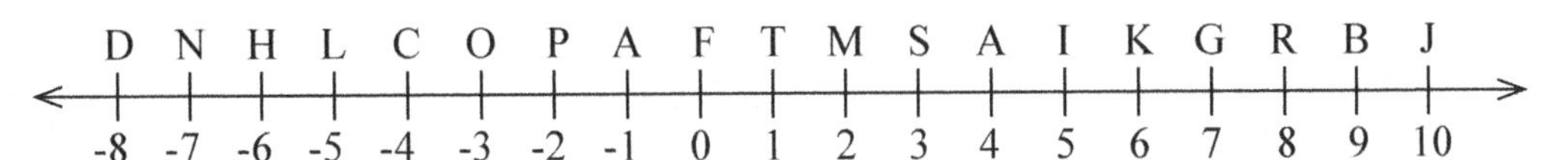

1. C   2. B, F, and J   3. S, M, and N   4. R, T, C, and D

Solve each equation and graph each solution on the number line. Be sure to label your answers. Also, place each solution in the answer column.

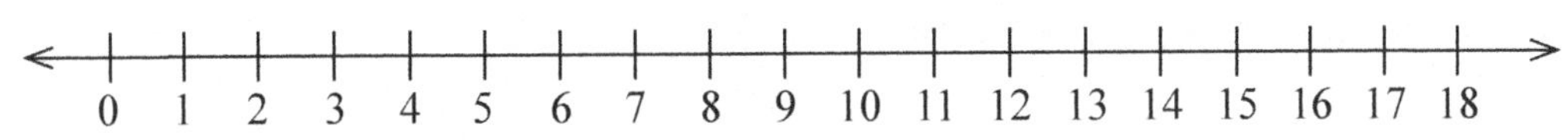

5. $n + 2 = 5$ 6. $x - 2 = 4$ 7. $3y = 15$ 8. $\frac{m}{2} = 4$

Use the coordinate system to find the ordered pair associated with each point.

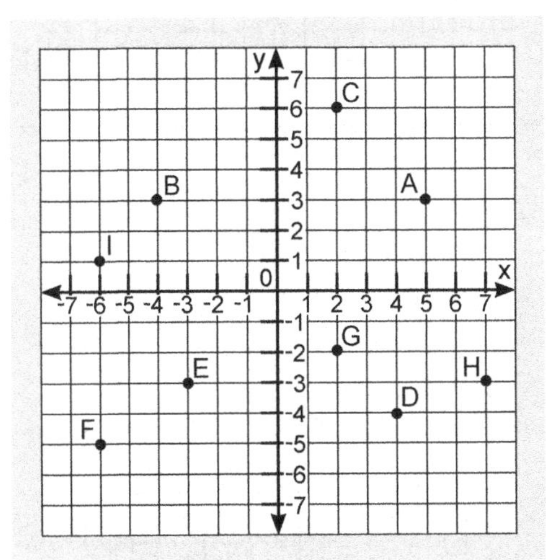

9. A

10. I

11. D

12. F

13. C

14. Find the slope of the line that passes through points A and G.

Use the coordinate system to find the ordered pair associated with each point.

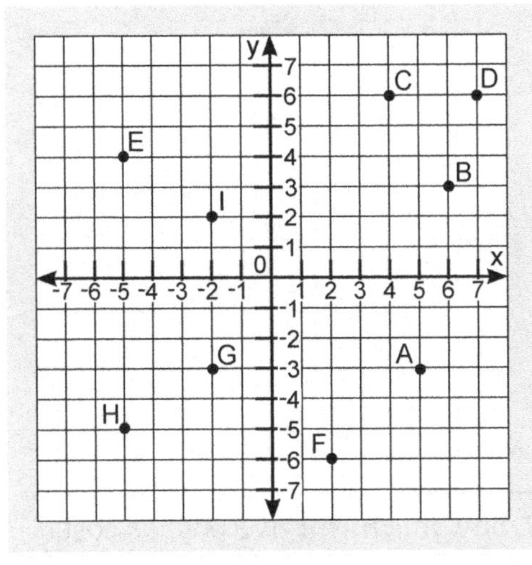

15. (6, 3)

16. (-2, 2)

17. (-5, -5)

18. (7, 6)

19. (5, -3)

20. Find the slope of the line that passes through points B and I.

1.
2.
3.
4.
5.
6.
7.
8.
9.
10.
11.
12.
13.
14.
15.
16.
17.
18.
19.
20.

Review Exercises

1. $-36 \div -6 =$
2. $-9 -6 + -3 =$
3. $-2 \times -3 \times -4 =$
4. $-7 - 9 =$
5. $-56 \div 8 =$
6. $(-2)^3 =$

Helpful Hints

The graph of a **linear equation** is always a line. A linear equation can have an infinite number of solutions, so to make a graph we select a few points and graph them, and then draw a line that connects them.

Example: Draw a graph of the solutions to the following equation.

$$y = x + 3$$

First, select four values for x and find the values for y. Start with x = 0 and make a chart like the one to the right.

x	y	
0	3	(0,3)
1	4	(1,4)
2	5	(2,5)
4	7	(4,7)

Next, plot the points and connect them with a line.

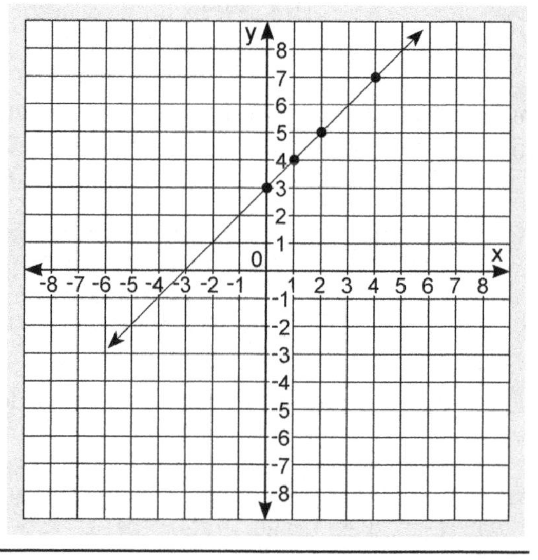

Make a table of 4 solutions. Graph the points. Connect them with a line

S1. $y = 2x + 1$

x	y

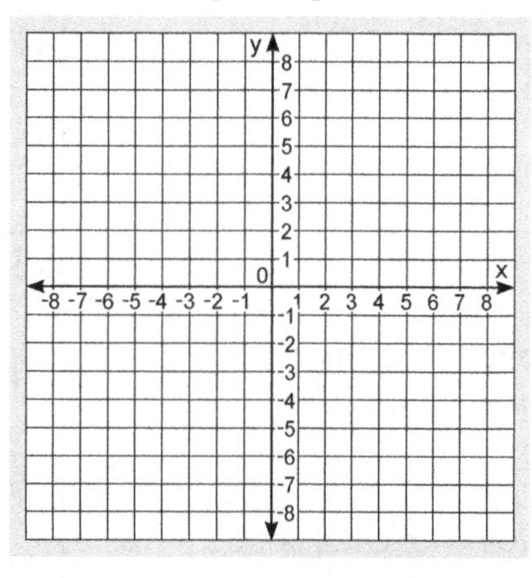

S2. $y = x - 3$

x	y

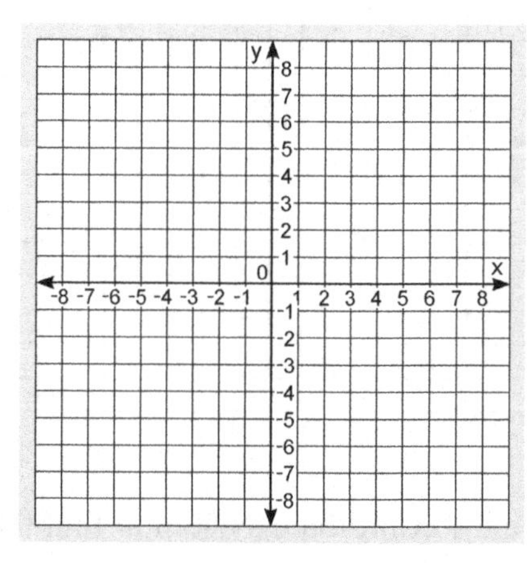

1. $y = 2x - 1$

x	y

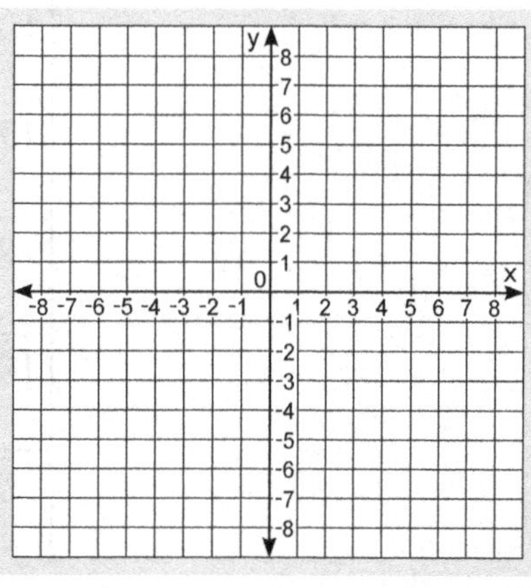

2. $y = \frac{x}{2}$

x	y

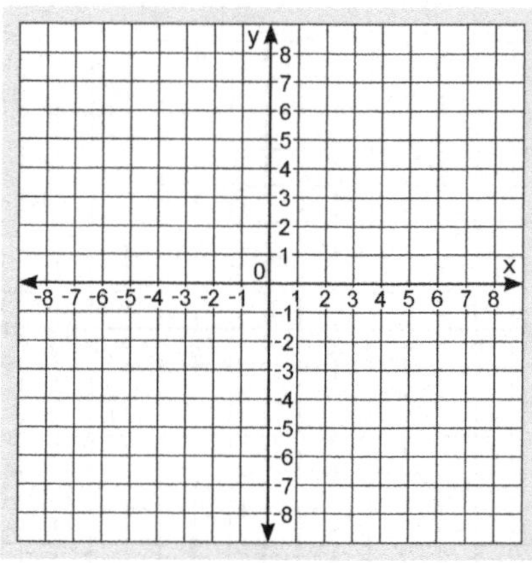

Problem Solving If three pounds of meat costs $3.60, how much will five pounds costs?

Review Exercises

1. Solve the proportion. $\frac{5}{6} = \frac{7}{n}$

2. Find 15% of 20.

3. 15 = 20% of what?

4. $-\frac{1}{3} + -\frac{3}{8} =$

5. $2 \times -1\frac{1}{2} =$

6. $-6.3 \div 3 =$

Helpful Hints

Use what you have learned to work the following problems.

Example: Draw a graph of the solutions to the following equation.

$y = \frac{x}{2} + 2$

x	y	
0	2	(0,2)
2	3	(2,3)
4	4	(4,4)
-2	1	(-2,1)

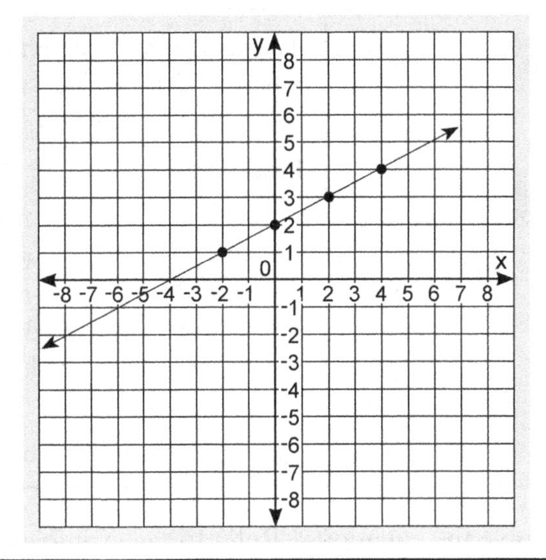

Make a table of 4 solutions. Graph the points. Connect them with a line

S1. $y = \frac{x}{3}$

x	y

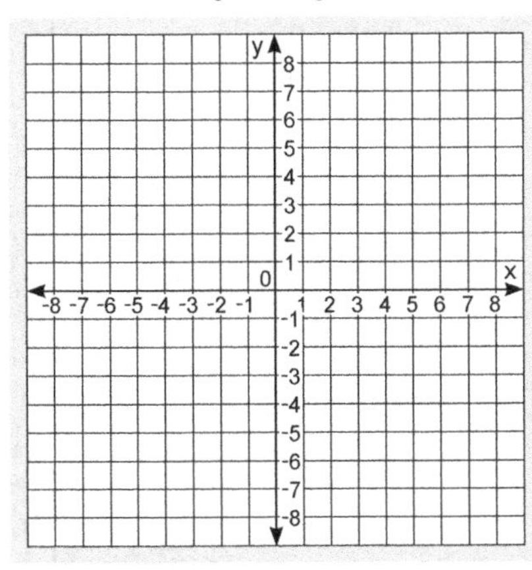

S2. $y = 2x + 3$

x	y

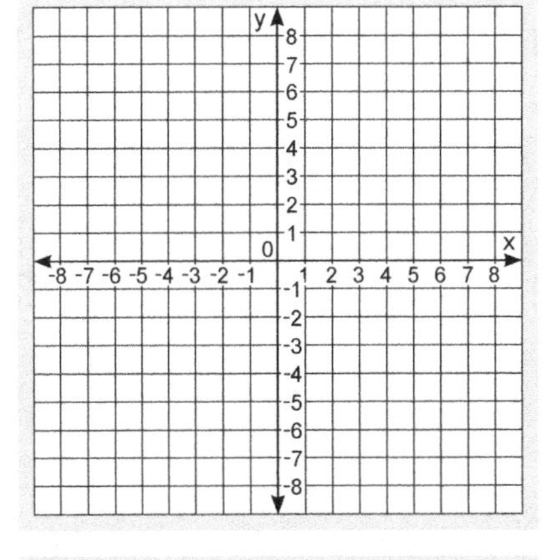

1. $y = \frac{x}{2} + 5$

x	y

2. $y = -2x$

x	y

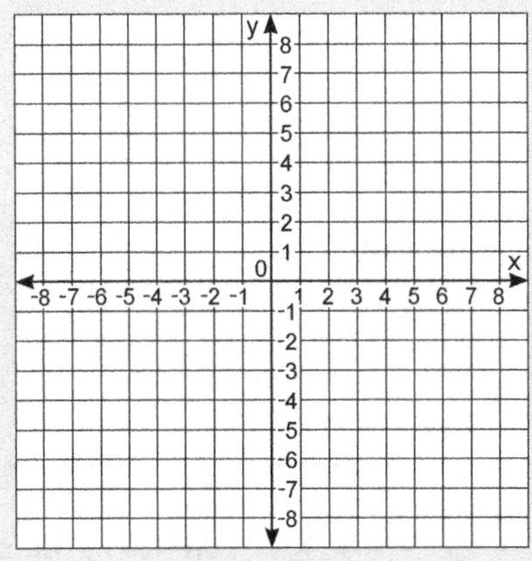

Problem Solving John has finished $\frac{4}{5}$ of the problems on a test. What percent has he finished?

Review Exercises

1. Write 1,720,000 in scientific notation.
2. Write .00000038 in scientific notation.
3. Write 1.963×10^8 as a conventional number.
4. Write 3.4×10^{-4} as a conventional number.
5. -9 - 7 - -6 =
6. -.34 + .53 =

Helpful Hints

The goal with any **equation** is to end up with the **variable** (letter), an **equal sign**, and the **answer**. You can add, subtract, multiply, and divide on each side of the equal sign with the same number, and won't change the solution.

Examples:

$x + 2 = 9$
$+ \quad -2 = -2$
$x = 7$ (circled)
Add -2 to both sides

$n - 6 = -5$
$+ \quad 6 = 6$
$n = 1$ (circled)
Add 6 to both sides.

$4n = 24$
$\frac{4n}{4} = \frac{24}{4}$
$n = 6$ (circled)
Divide both sides by 4.

$\frac{x}{6} = 4$
$\frac{6}{1} \times \frac{x}{6} = 4 \times 6$
$x = 24$ (circled)
Multiply both sides by 6.

Check your work by substituting your answer in the original equation.

Solve the equations. Refer to the examples above.

S1. $x + 3 = 8$

S2. $3n = 96$

1. $n - 5 = -8$

2. $\frac{n}{5} = 8$

3. $n + 6 = -7$

4. $5n = -25$

5. $n + -6 = 7$

6. $\frac{n}{4} = -3$

7. $x + 23 = 57$

8. $15n = 60$

9. $n - -6 = -5$

10. $n + 12 = -15$

1.
2.
3.
4.
5.
6.
7.
8.
9.
10.
Score

Problem Solving

What is the slope of a line that passes through the points (6, 1) and (9, 8)?

Review Exercises

1. $\frac{3^2 + 2^2 + 7}{2} =$

2. $3 + 7 \times 2 + 6 =$

3. $3 \times (7^2 - 15) =$

4. $n + 5 = -3$

5. $3n + 18 =$

6. $\frac{n}{3} = 7$

Helpful Hints

Be careful with negative signs when solving equations.

Examples:

$$\begin{aligned} -x + 7 &= -9 \\ + \quad -7 &= -7 \\ \hline -x &= -16 \end{aligned}$$

If $-x = -16$, then $x = 16$

$$-3n = 18$$
$$\frac{-3n}{-3} = \frac{18}{-3}$$

Divide both sides by -3.

$n = -6$

$$\frac{n}{-5} = 7$$
$$\frac{-5}{1} \times \frac{n}{-5} = 7 \times -5$$

Multiply both sides by -5.

$n = -35$

* Remember to check your work by substituting your answer in the original equation.

Solve the equations. Refer to the examples above.

S1. $-x + 7 = -5$

S2. $-4n = -12$

1. $-n - 6 = 8$

2. $\frac{n}{-2} = 6$

3. $-x + -7 = 2$

4. $-3n = 15$

5. $n - 6 = 12$

6. $5n = -30$

7. $-n - 6 = -8$

8. $\frac{n}{-4} = -5$

9. $3n = -45$

10. $-n + -6 = -20$

Answers
1.
2.
3.
4.
5.
6.
7.
8.
9.
10.
Score

Problem Solving

Write the ratio 18 to 8 as a fraction reduced to lowest terms.

Review Exercises

1. $-7 - -9 + 6 - 7 =$
2. $3 \times -2 \times 4 \times -3 =$
3. $\dfrac{-64 \div 8}{24 \div -6} =$
4. Write 210,000 in scientific notation.
5. Write .00316 in scientific notation.
6. $(-2)^4 =$

Helpful Hints

Some equations require two steps.

Examples:

$$\begin{aligned} 2x - 5 &= 71 \\ +\ 5 &= 5 \\ \frac{2x}{2} &= \frac{76}{2} \\ x &= \boxed{38} \end{aligned}$$

Add 5 to both sides. Divide both sides by 2.

$$\begin{aligned} \frac{n}{5} + 3 &= 8 \\ +\ -3 &= -3 \\ \frac{5}{1} \times \frac{n}{5} &= 5 \times 5 \\ n &= \boxed{25} \end{aligned}$$

Add 3 to both sides. Multiply both sides by 5.

$$\begin{aligned} -3n - 4 &= 11 \\ +\ 4 &= 4 \\ \frac{-3n}{-3} &= \frac{15}{-3} \\ n &= \boxed{-5} \end{aligned}$$

Add 4 to both sides. Divide both sides by -3.

* Remember to check your work by substituting your answer in the original equation.

S1. $3x - 5 = 16$

S2. $\frac{x}{2} + 2 = 4$

1. $7x + 3 = -4$
2. $-14n - 7 = 49$
3. $2n + 45 = 15$
4. $\frac{n}{5} + -6 = 9$
5. $4x - 10 = 38$
6. $-2m + 9 = 7$
7. $35x + 12 = 82$
8. $\frac{m}{5} - 7 = 3$
9. $3x - 12 = 18$
10. $5x + 2 = -13$

1.
2.
3.
4.
5.
6.
7.
8.
9.
10.
Score

Problem Solving

Six students were absent Monday at Jefferson School. If this was 3% of the total enrollment, how many students are enrolled at Jefferson School?

Review Exercises

1. $-\frac{2}{5} + \frac{1}{2} =$
2. $-\frac{2}{5} + -\frac{2}{5} =$
3. $\frac{2}{3} \div -\frac{1}{2} =$
4. $1\frac{1}{2} \times -2 =$
5. $.2 \times -3.2 =$
6. $-.6 + -.5 =$

Helpful Hints

Sometimes the **distributive property** can be used to solve equations.

Examples:

$2(x + 7) = 30$

First use the distributive property.

$2(x + 7) = 30$

$2x + 14 = 30$

$+ \quad -14 = -14$ Add -14 to both sides.

$\frac{2x}{2} = \frac{16}{2}$ Divide both sides by 2.

$x = 8$

$3(4x - 3) = -33$

First use the distributive property.

$3(4x - 3) = -33$

$12x - 9 = -33$

$+ \quad 9 = 9$ Add 9 to both sides.

$\frac{12x}{12} = \frac{-24}{12}$ Divide both sides by 12.

$x = -2$

* Remember to check your answers.

Solve the following equations. Use the distributive property when necessary.

S1. $5(m + 6) = 45$

S2. $\frac{x}{5} + -6 = 3$

1. $3(m - 2) = 18$
2. $3x + 7 = -2$
3. $4m - 9 = 31$
4. $2(m + -2) = -10$
5. $-6x + 2 = -28$
6. $-x + 8 = 12$
7. $\frac{x}{2} + 3 = -2$
8. $2x + 1 = -13$
9. $5x - 3 = -18$
10. $4(x + 2) = 48$

1.
2.
3.
4.
5.
6.
7.
8.
9.
10.
Score

Problem Solving

Find the greatest common factor of 42 and 56.

Review Exercises

1. $6^3 =$
2. $7^0 =$
3. $9^1 =$
4. $\sqrt{36} + \sqrt{49} =$
5. $2^3 + \sqrt{16}$
6. $33 + 5^3 =$

Helpful Hints

Sometimes there are variables on both sides of the equal sign.

Examples:

$$
\begin{aligned}
5x - 6 &= 2x + 9 \\
+\ -2x \quad &= -2x \quad \text{Add -2x to both sides.} \\
3x - 6 &= 9 \\
+\ 6 &= 6 \\
\frac{3x}{3} &= \frac{3}{3} \quad \text{Divide both sides by 3.} \\
x &= \boxed{1}
\end{aligned}
$$

$$
\begin{aligned}
-6x + 12 &= 4x - 8 \\
+6x \quad &= 6x \quad \text{Add 6x to both sides.} \\
12 &= 10x - 8 \\
+\ 8 &= 8 \\
\frac{20}{10} &= \frac{10x}{10} \quad \text{Divide both sides by 10.} \\
\boxed{2} &= x
\end{aligned}
$$

S1. $4x + 3 = 2x + 1$

S2. $5x + 1 = 2x + 10$

1. $4x - 12 = 2x + 2$
2. $x - 2 = 2x - 4$
3. $7x - 16 = x + 8$
4. $7x - 1 = 15 + 3x$
5. $-2x + 8 = 4x - 10$
6. $3x + 6 = x + 8$
7. $3x - 5 = x - 7$
8. $3x + 5 = x + 13$
9. $2x + 6 = -x + 12$
10. $5x - 6 = 2x + 12$

Answers
1.
2.
3.
4.
5.
6.
7.
8.
9.
10.
Score

Problem Solving

Susan had 30 apples and used six of them to make a pie. What percent of the apples did she use to make the pie?

Review Exercises

1. Solve the proportion. $\frac{n}{4} = \frac{25}{5}$

2. Is $\frac{4}{7} = \frac{3}{5}$ a proportion? Why?

3. Write 16 to 6 as a fraction reduced to lowest terms.

4. Find 20% of 300.

5. Six is what % of 24?

6. 7 = 20% of what?

Helpful Hints

Use what you have learned to solve the following equations.
* If necessary, refer to the previous Helpful Hints sections.
* Check your answers by substituting them in the original equation.

Solve the following equations. Use the distributive property when necessary.

S1. $7n = 28$

S2. $\frac{n}{5} = 3$

1. $2x + 6 = 2$

2. $3x + 7 = -2$

3. $4m - 9 = 31$

4. $4(x + 3) = -8$

5. $3(n - 2) = 30$

6. $3x + 6 = x + 8$

7. $4x - 12 = 2x + 2$

8. $-5x + 2 = -13$

9. $\frac{x}{5} - 2 = -7$

10. $3(x + 4) = -18$

1.
2.
3.
4.
5.
6.
7.
8.
9.
10.
Score

Problem Solving

Jeff has a marble collection. The ratio of red marbles to blue marbles is three to two. If he has 12 red marbles, how many blue marbles does he have? (Use a proportion.)

Review Exercises

1. List all the factors of 48.
2. What is the GCF of 16 and 24?
3. What is the LCM of 6 and 10?
4. $3n = 15$, $n =$
5. $\frac{n}{2} = 10$, $n =$
6. $-x = 5$, $x =$

Helpful Hints

To solve **algebra word problems**, it is necessary to translate words into **algebraic expressions** containing a **variable**. A **variable** is a letter that represents a number. Here are some examples:

Three more than a number → $x + 3$
Twice a number → $2x$
The quotient of x and five → $\frac{x}{5}$
Four less than a number → $x - 4$
Seven times a number → $7x$
A number decreased by six → $x - 6$
Seven less than three times a number → $3x - 7$
Twice a number less nine is equal to 15 → $2x - 9 = 15$
The difference between three times a number and eight equals 12 → $3x - 8 = 12$
The sum of a number and -9 is 24 → $x + -9 = 24$
Three times a number less six equals twice the number plus 15 → $3x - 6 = 2x + 15$
Twice the sum of n and five → $2(n + 5)$
The difference between four times x and 15 equals twice the number → $4x - 15 = 2x$

Translate each of the following into an equation.

S1. Seven less than twice a number is 12.

S2. Two more than three times a number equals 30.

1. The sum of twice a number and five is 14.
2. The difference between four times a number and six is 10.
3. Twelve is five less than four times a number.
4. One-third times a number less four equals twice the number added to eight.
5. Twice the sum of a number and two equals 10.
6. The difference between five times a number and three is 17.
7. Twice a number decreased by six is 15.
8. Two less than three times a number is seven more than twice the number.
9. Four more than a number equals the sum of seven and -12.
10. A number divided by five is 25.

Answers
1.
2.
3.
4.
5.
6.
7.
8.
9.
10.
Score

Problem Solving

If a car can travel 65 miles per hour, how far can it travel in 3.5 hours?

Review Exercises

1. $x + 2 = 9$
   $x =$

2. $n + -3 = -7$
   $n =$

3. $3n = 36$

4. $-5n = -25$

5. $\frac{n}{3} = 5$
   $n =$

6. $2x + 1 = 7$

Helpful Hints

Algebra word problems must be translated into an **equation** and solved.

Example:

Six times a number less two equals four times the number added to 10.
First translate and then solve.

$$6x - 2 = 4x + 10$$
$$+ -4x \qquad -4x$$ Add -4x to both sides.
$$2x - 2 = 10$$
$$2 = 2$$ Add 2 to both sides.
$$\frac{2x}{2} = \frac{12}{2}$$ Divide both sides by 2.
$$x = 6$$ The number is 6.

Translate each of the following into an equation and solve.

S1. Six less than twice a number is 16. Find the number.

S2. The difference between three times a number and 8 is 28. Find the number.

1. Five less than twice a number is 67. Find the number.

2. Four times a number decreased by five is -17. Find the number.

3. Four times a number less six is eight more than two times the number. Find the number.

4. Eight more then one-half a number is 10. Find the number.

5. The difference between four times a number and two is 10.

1.
2.
3.
4.
5.
Score

Problem Solving

A doctor's annual income is $150,000. What is his average monthly income?

Review Exercises

1. Write 3.61×10^{-7} as a conventional number.

2. Write .00000127 in scientific notation.

3. Write 729,000,000 in scientific notation.

Helpful Hints

Remember these steps when solving algebra word problems.

1. Read the problem very carefully.
2. Write an equation.
3. Solve the equation and find the answer.
4. Check your answer to be sure it makes sense.

Example: John is twice as old as Susan. The sum of their ages is 42. What is each of their ages?

Let x = Susan's age   2x = John's age

$x + 2x = 42$ Susan's age is x = (14.)
$3x = 42$ John's age is 2x = (28.)
$x = 14$ The sum is 42.

Solve the algebra word problems.

S1. Amir is six years older than Kevin. The sum of their ages is 30. Find the age of each.

S2. A board 44 inches long is cut into two pieces. The long piece is three times the length of the short piece. What is the length of each piece.

1. Bob and Bill together earn $66. Bill earned $6 more than twice as much as Bob. How much did each earn?

2. Steve worked Monday and Tuesday and earned a total of $212. He earned $30 more on Tuesday than he did on Monday. How much did Steve earn each day?

3. Five times Bob's age plus six equals three times his age plus 30. What is Bob's age?

4. Sixty dollars less than three times Susan's weekly salary is equal to 360 dollars. What is Susan's weekly salary?

5. Twice John's age less 12 is 48. What is John's age?

1.
2.
3.
4.
5.
Score

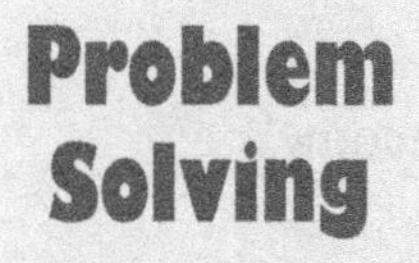

A student has test scores of 90, 96, 84, and 86. What was his average score?

Review Exercises

Solve each equation.

1. $2x + 7 = -15$
2. $5x + 6 = 106$
3. $\frac{n}{4} + 2 = 13$
4. $3(n + 6) = -9$
5. $5x + 3 = 7x + -3$
6. $3x + 2x = 55$

Helpful Hints

*Remember:
1. Read the problem carefully.
2. Write an equation.
3. Solve the equation and find the answer.
4. Check your answer to be sure it makes sense.

Solve each algebra word problem.

S1. Five more than six times a number is equal to 48 less 7. Find the number.

S2. Steve weighs 50 pounds more than Bart. Their combined weight is 270 pounds. What is each of their weights?

1. The sum of three times a number and 15 is -12. Find the number.
2. Eight more than six times a number is 20 more than four times the number. Find the number.
3. The sum of five and a number is -19. Find the number.
4. Roy is three times as old as Ellen. The sum of their ages is 44 years. What are each of their ages?
5. Six more than two times a number is six less than six times the number. Find the number.

1.
2.
3.
4.
5.
Score

Problem Solving

A plane travelled 2,100 miles in 3.5 hours. What was the plane's average speed per hour?

Reviewing Equations and Algebra Word Problems

For 1 - 12, solve each equation. Be sure to show all work.

1. $x + 5 = -2$
2. $3n = 39$
3. $\frac{n}{7} = 8$
4. $5n + 2 = 17$
5. $3n - 6 = -21$
6. $\frac{n}{3} - 6 = -12$
7. $3(n + 2) = -15$
8. $5(x - 4) = 55$
9. $2x + 4 = 4x - 12$
10. $5x - 3 = 3x + 13$
11. $3x + 4x = -77$
12. $\frac{n}{-3} + 2 = -5$

For 13 - 20, solve each algebra word problem

13. Twice a number less three is 21. Find the number.
14. Eight more than five times a number is -17. Find the number.
15. The difference between five times a number and six is 24. Find the number.
16. Seven more than twice a number is five less than four times the number. Find the number.
17. Ann has twice as much money as Sue. Together they have $66. How much does each have?
18. Bill is eight years older than Ron. The sum of their ages is 64 years. How old is each of them?
19. Four times a number decreased by six equals -14. Find the number.
20. Four more than one-third of a number is 10. Find the number.

1.
2.
3.
4.
5.
6.
7.
8.
9.
10.
11.
12.
13.
14.
15.
16.
17.
18.
19.
20.

Review Exercises

1. List the first seven multiples of 8.
2. List all factors of 60.
3. What is the GCF of 100 and 40?
4. Write .000006 in scientific notation.
5. Write 2,100,000 in scientific notation.
6. Write 2.1×10^{-3} as a conventional number.

Helpful Hints

Probability tells what chance, or how likely it is for an event to occur. Probability can be written as a fraction.

$$\textbf{Probability} = \frac{\text{number of ways a certain outcome can occur}}{\text{number of possible outcomes}}$$

Examples: If you toss a coin, what is the probability that it will show heads?

$$\frac{\text{1 - heads is one outcome}}{\text{2 - there are two possible outcomes, heads or tails}}$$

The probability is 1 out of 2.

There are six marbles in a jar. Three are red, two are blue, and one is green. What is the probability that you will draw a blue one without looking?

$$\frac{\text{2 - blue marbles}}{\text{6 - marbles in the jar}}$$

The probability is 2 out of 6, or simplified, 1 out of 3.

Use the information below to answer the following questions.

There are 3 red marbles, 6 blue marbles, 2 black marbles, and 1 green marble in a can. Find the probability of each of the following.

S1. A red marble.

S2. A blue or green marble.

1. A black marble.
2. A green marble.
3. A blue or red marble.
4. Not a black marble.
5. Not a red marble.
6. Not a green or blue marble.
7. A green, red, or blue marble.
8. Not a blue marble.
9. A green, red, or black marble.
10. Not a blue or black marble.

1.
2.
3.
4.
5.
6.
7.
8.
9.
10.
Score

Problem Solving

Four times a number less five is -17. Find the number.

Review Exercises

Solve each of the following equations.

1. $3x + 2 = -28$
2. $\frac{x}{5} - 6 = -11$
3. $4(n + 3) = -28$
4. $2x + 10 = 4x + 2$
5. $3x + 2x = 75$
6. $7x - 3 = 60$

Helpful Hints

Use what you have learned to solve the following questions.
Example: What is the probability of the spinner landing on the 1 or the 3?
2 out of 8 or, simplified, 1 out of 4.

Use the spinner to find the probability for each of the following.
Find the probability of spinning once and landing on each of the following.

S1. a three
S2. an even number

1. a seven
2. not a five
3. an odd number
4. a number less than five
5. a number greater than six
6. a nine
7. a one or an eight
8. an even number or a five
9. a number greater than three
10. a number which is a factor of six

1.
2.
3.
4.
5.
6.
7.
8.
9.
10.
Score

Problem Solving

If five pounds of beef cost $9, how many pounds can be bought with $36?

Review Exercises

1. Change .3 to a percent.
2. Change .03 to a percent.
3. Change $\frac{3}{5}$ to a percent.
4. Find 4% of 50.
5. Fifteen is what % of 60?
6. 4 = 20% of what?

Helpful Hints

Statistics involves gathering and recording **data**. Number facts about events or objects are called data. The **range** is the difference between the greatest number and the least number in a list of data. The **mode** is the number which appears the most in a list of data.

Example: Find the range and mode for the list of data.
12, 10, 1, 7, 4, 7, 5
First, list the numbers from least to greatest.
1, 4, 5, 7, 7, 10, 12
The range is 12 - 1 = 11.
The mode is 7, which appears the most.

Arrange the data in order from least to greatest, then find the range and mode.

S1. 7, 4, 1, 8, 2, 5, 4

S2. 6, 2, 7, 6, 8, 2, 5, 6, 3

1. 7, 4, 8, 2, 4, 7, 7
2. 25, 17, 30, 39, 16, 24, 30
3. 1, 3, 6, 3, 4, 6, 11, 9
4. 1, 6, 17, 8, 9, 20, 9
5. 7, 3, 1, 3, 1, 3, 8, 4
6. 3, 14, 8, 6, 11, 8, 14, 8
7. 1, 10, 2, 9, 3, 8, 2, 7
8. 85, 91, 90, 86, 91, 87
9. 1, 10, 2 9, 2, 7, 2, 8
10. 20, 2, 19, 1, 2, 16, 3

1.
2.
3.
4.
5.
6.
7.
8.
9.
10.
Score

Problem Solving

If three cans of juice cost $1.14, what is the cost of one can?

Review Exercises

1. Write 16 to 10 as a fraction reduced to lowest terms.

2. Is $\frac{9}{11} = \frac{7}{8}$ a proportion? Why?

3. Solve the proportion. $\frac{4}{n} = \frac{9}{45}$

4. Write 1,280,000 in scientific notation.

5. Write .0000962 in scientific notation.

6. Write 6.2×10^{-5} as a conventional number.

Helpful Hints

The **mean** of a list of data is found by adding all the items in the list and then dividing by the number of items.

The **median** is the middle number, when the list of data is arranged from least to greatest.

Example: Find the mean and median for the list of data.
1, 2, 5, 6, 6

Median = (5) Mean = $\frac{1+2+5+6+6}{5} = \frac{20}{5} = $ (4)

Arrange the data in order from least to greatest, then find the mean and median.

S1. 1, 5, 2, 4, 3

S2. 6, 1, 7, 4, 2, 6, 2

1. 2, 7, 1, 4, 1
2. 1, 5, 7, 1, 2, 2, 3
3. 5, 25, 10, 20, 15
4. 1, 1, 1, 3, 3, 3, 4, 1, 1
5. 8, 5, 2, 9, 3, 6, 9
6. 126, 136, 110
7. 7, 3, 4, 2, 4
8. 3, 1, 4, 7, 5
9. 2, 10, 4, 8, 1
10. 50, 70, 30

1.
2.
3.
4.
5.
6.
7.
8.
9.
10.
Score

Problem Solving

In a class of 40 students, 20% of them received A's.
How many students did not receive A's?

Review Exercises

1. $3 + 4 \times 5 - 2 =$
2. $3(8 + 2) - 4^2 =$
3. $(15 - 8) + 64 \div 2^3 =$
4. $7 \times 4 - 39 \div 3 =$
5. $6\ [(3 + 4) \times 2] =$
6. $3(-2 + 4) + 5 =$

Helpful Hints

Use what you have learned to answer the following questions.

* If necessary, refer to the two previous pages.

Arrange the data in order from least to greatest, then answer the questions.

2, 8, 6, 2, 7

S1. What is the range?

S2. What is the mode?

1. What is the mean?
2. What is the median?

1, 9, 2, 7, 2, 3, 4

3. What is the median?
4. What is the mode?
5. What is the range?
6. What is the mean?

2, 11, 8, 6, 1, 2, 5

7. What is the range?
8. What is the mode?
9. What is the mean?
10. What is the median?

1.
2.
3.
4.
5.
6.
7.
8.
9.
10.
Score

Problem Solving

Light travels at a speed of 1.86×10^5 miles per second. Write the speed as a conventional number.

Reviewing Probability and Statistics

There are four green marbles, three red marbles, two white marbles, and one blue marble in a can. What is the probability for each of the following?

1. a red marble
2. a green marble
3. a green or blue marble
4. not a red marble
5. a green, red, or blue marble
6. not a green marble

Use the spinner to find the probability of spinning once and landing on each of the following.

7. a five
8. an odd number
9. a number greater than three
10. a one or a three
11. a number less than five
12. a one or a six

Arrange the data in order from least to greatest, then answer the questions.

2, 5, 4, 10, 4

13. What is the range?
14. What is the mode?
15. What is the mean?
16. What is the median?

2, 5, 2, 1, 3, 7, 8

17. What is the mode?
18. What is the mean?
19. What is the range?
20. What is the median?

1.
2.
3.
4.
5.
6.
7.
8.
9.
10.
11.
12.
13.
14.
15.
16.
17.
18.
19.
20.

Final Review - All Pre-Algebra Concepts

For 1 - 3, use the following sets to find the answers.

A = {1,2,3,4,5}, B = {2,3,4,6,8}, C = {0,1,2,4,5,9}

1. Find $A \cap B$
2. Find $B \cup C$
3. Find $A \cap C$
4. $-9 + 12 =$
5. $-16 - 7 =$
6. $-12 \times -3 =$
7. $-24 \div -3 =$
8. $.21 + -.76 =$
9. $-\frac{2}{5} + -\frac{1}{2} =$
10. $5^3 =$
11. $\sqrt{49}$
12. $3^3 + \sqrt{36} =$
13. $6 + 7 \times 3 - 5 =$
14. $3^2(3 + 4) + 5 =$
15. $\frac{4^2 + 12}{5 + 3(2+1)} =$
16. $2[(5 + 7) \div 3 + 6] =$
17. What property is illustrated below?

    $5 + 6 = 6 + 5$
18. What property is illustrated below?

    $7(6 + 5) = 7(6) + 7(5)$
19. Write 1,280,000,000 in scientific notation.
20. Write .00000653 in scientific notation.

1.
2.
3.
4.
5.
6.
7.
8.
9.
10.
11.
12.
13.
14.
15.
16.
17.
18.
19.
20.

Final Review - All Pre-Algebra Concepts

21. Write 6.09×10^7 as a conventional number.

22. Write $7.62 \times 10\text{-}6$ as a conventional number.

23. Write 18 to 10 as a fraction reduced to lowest terms.

24. Is $\frac{7}{8} = \frac{3}{4}$ a proportion? Why?

25. Solve the proportion.

$$\frac{7}{n} = \frac{3}{9}$$

26. The ratio of red marbles to blue marbles is five to two. If there are 15 red marbles, how many blue marbles are there?

27. Find 5% of 80.

28. Six is what percent of 24?

29. 8 = 25% of what?

30. On a test with 30 questions, a student got 80% correct. How many questions did he get correct?

31. There are 35 fish in an aquarium. If 14 of them are goldfish, what percent of them are goldfish?

32. Six students get A's on a test. This is 20% of the class. How many are there in the class?

33. Find all the factors of 40.

34. Find the GCF of 60 and 40.

35. Find the first seven multiples of six.

36. Find the LCM of 8 and 12.

Use the number line to state the coordinates of the given points.

A	D	I	N	J	P	E	T	B	S	K	R	F	L	Q	C	M	H	G
-8	-7	-6	-5	-4	-3	-2	-1	0	1	2	3	4	5	6	7	8	9	10

37. A, E, B

38. N, H, T, D

39. I, F, P

40. H, M, S

21.

22.

23.

24.

25.

26.

27.

28.

29.

30.

31.

32.

33.

34.

35.

36.

37.

38.

39.

40.

Final Review - All Pre-Algebra Concepts

For 41 - 50, use the coordinate system to answer each question.

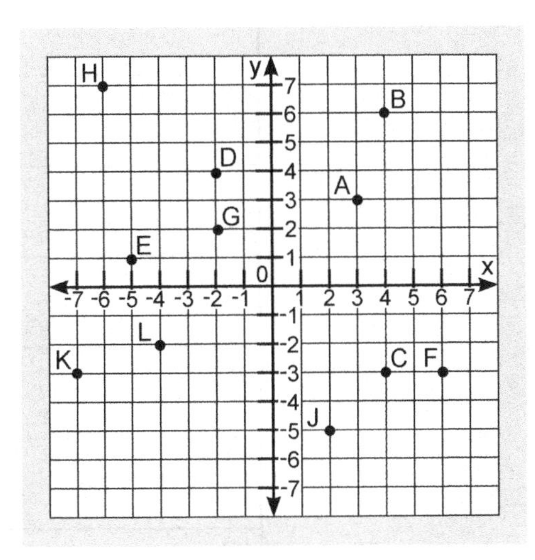

For 41-45, find the ordered pair associated with each point.

41. F 42. E

43. C 44. L

45. B

For 46-52, find the point associated with each ordered pair.

46. (-6, 7) 47. (-2, 2) 48. (-4, -2) 49. (2, -5) 50. (-7, -3)

51. Find the slope of the line that passes through the points (1, 3) and (4, 5).

52. Find the slope of the line that passes through the points (-2, 5) and (6, 8).

For 53 through 54, make a table of 4 solutions and graph the points. Connect them with a line.

53. $y = x + 4$

x	y

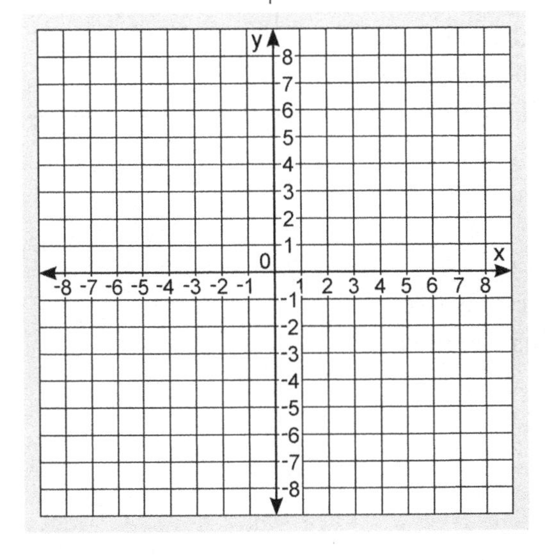

54. $y = 2x + 2$

x	y

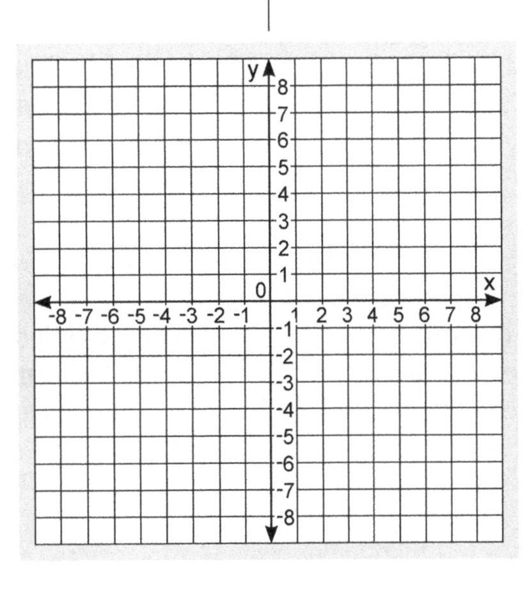

41.
42.
43.
44.
45.
46.
47.
48.
49.
50.
51.
52.
53.
54.

Final Review - All Pre-Algebra Concepts

Solve each equation and word problem.

55. $x + 3 = 12$

56. $3n = -45$

57. $\frac{n}{6} = 3$

58. $-5n = 15$

59. $2x + 3 = 15$

60. $5x - 2 = -17$

61. $\frac{n}{3} + -4 = 4$

62. $3(x + 4) = 24$

63. $3(x + 4) = -6$

64. $2x + 12 = 4x + 10$

65. Two more than three times a number is 29. Find the number.

66. Twice a number, less seven, is 17. Find the number.

67. A number divided by five, less six, is four. Find the number.

68. Sue has three times as much money as Jane. Together they have 64 dollars. How much does each have?

69. Al is seven years older than Maria. The sum of their ages is 51. What is each of their ages?

70. Six times Glen's age plus two equals four times his age plus 20. Find his age.

55.
56.
57.
58.
59.
60.
61.
62.
63.
64.
65.
66.
67.
68.
69.
70.

Final Review - All Pre-Algebra Concepts

Use the following information to answer 71 - 74.

There are 6 green marbles, 5 red marbles, 4 white marbles, and 1 blue marble in a can. What is the probability for each of the following?

71. a red marble

72. a green marble

73. a green or blue marble

74. not a red marble

Use the spinner to find the probability of spinning once and landing on each of the following.

75. a seven.

76. an even number.

77. a number greater than four.

78. a one, a three, or a five.

Arrange the data in order from least to greatest, then answer the questions.

2, 7, 3, 10, 3

79. What is the range?

80. What is the mode?

81. What is the mean?

82. What is the median?

4, 10, 4, 2, 6, 14, 16

83. What is the mode?

84. What is the mean?

85. What is the range?

86. What is the median?

71.
72.
73.
74.
75.
76.
77.
78.
79.
80.
81.
82.
83.
84.
85.
86.

Final Review - All Pre-Algebra Concepts

Solve each of the following problems.

87. $\begin{array}{r} \frac{7}{8} \\ + \frac{3}{8} \\ \hline \end{array}$

88. $\begin{array}{r} 7\frac{1}{4} \\ - 3\frac{3}{4} \\ \hline \end{array}$

89. $\begin{array}{r} 3\frac{3}{5} \\ + 2\frac{1}{10} \\ \hline \end{array}$

90. $\frac{5}{8} \times 3\frac{1}{5} =$

91. $2\frac{1}{2} \times 3\frac{1}{2} =$

92. $2\frac{1}{3} \div \frac{1}{2} =$

93. $5\frac{1}{2} \div 1\frac{1}{2} =$

94. $.6 + 7.62 + 5.2 + 6 =$

95. $6.3 - 1.275 =$

96. $72 - 1.68 =$

97. $\begin{array}{r} 2.19 \\ \times\ 7 \\ \hline \end{array}$

98. $\begin{array}{r} .36 \\ \times\ 1.2 \\ \hline \end{array}$

99. $5\overline{)6.7}$

100. $.15\overline{).0045}$

87.
88.
89.
90.
91.
92.
93.
94.
95.
96.
97.
98.
99.
100.

Answer Key

Geometry—Solutions

Page 12

Review Exercises

1. 1,085
2. 636
3. 1,482
4. 184
5. 584
6. 3,152

S1. answers vary
S2. answers vary
1. answers vary
2. answers vary
3. D, E, F
4. $\overleftrightarrow{AC}$, $\overleftrightarrow{BC}$
5. $\overleftrightarrow{FE}$, $\overleftrightarrow{FB}$
6. $\overrightarrow{AB}$, $\overrightarrow{BC}$
7. $\overline{FE}$, $\overline{ED}$, $\overline{FD}$
8. answers vary
9. answers vary
10. Point E

Problem Solving: $6,252

Page 13

Review Exercises

1. 1,087
2. 109
3. 14,959
4. 1,058
5. 1,180
6. 4,213

S1. answers vary
S2. answers vary
1. answers vary
2. answers vary
3. G, I, D, C
4. answers vary
5. $\overleftrightarrow{JH}$, $\overleftrightarrow{IH}$
6. $\overrightarrow{EA}$, $\overrightarrow{DA}$
7. answers vary
8. answers vary
9. answers vary
10. D

Problem Solving: 18,200 cars

Page 14

Review Exercises

1. 212
2. 123 r2
3. 296 r4
4. 4,606
5. 1,397
6. 4,343

S1. $\overleftrightarrow{DK}$, $\overleftrightarrow{EJ}$
S2. answers vary
1. answers vary
2. answers vary
3. answers vary
4. answers vary
5. answers vary
6. answers vary
7. answers vary
8. answers vary
9. answers vary
10. $\angle$H, $\angle$IHJ

Problem Solving: 87

Page 15

Review Exercises

1. answers vary
2. answers vary
3. answers vary
4. 7,248
5. 6,445
6. 121

S1. answers vary
S2. $\overleftrightarrow{BM}$, $\overleftrightarrow{CF}$
1. answers vary
2. answers vary
3. answers vary
4. answers vary
5. answers vary
6. answers vary
7. answers vary
8. answers vary
9. answers vary
10. $\angle$L, $\angle$HLB

Problem Solving: $154

Page 16

Review Exercises

1. 1,147
2. answers vary
3. 735
4. answers vary
5. 3,380
6. 17,220

S1. answers vary
S2. answers vary
1. answers vary
2. answers vary
3. acute
4. obtuse
5. right
6. straight
7. answers vary
8. answers vary
9. answers vary
10. answers vary

Problem Solving:
55 miles per hour

Page 17

Review Exercises

1. 1,631
2. 4,249
3. 648
4. 19,200
5. 146,730
6. 1,001

S1. answers vary
S2. answers vary
1. answers vary
2. answers vary
3. acute
4. obtuse
5. right
6. 60°
7. straight
8. acute
9. obtuse
10. 150°

Problem Solving:
27 are empty

Page 18

Review Exercises

1. acute
2. obtuse
3. answers vary
4. answers vary
5. answers vary
6. 8,234

S1. acute, 20°
S2. obtuse, 110°
1. right, 90°
2. obtuse, 160°
3. acute, 20°
4. acute, 70°
5. obtuse, 130°
6. acute, 50°
7. straight, 180°
8. obtuse, 160°
9. right, 90°
10. obtuse, 130°

Problem Solving: 21 students

Page 19

Review Exercises

1. answers vary
2. answers vary
3. answers vary
4. answers vary
5. 1,691
6. 4,428

S1. 20°, acute
S2. 65°, acute
1. 100°, obtuse
2. 145°, obtuse
3. 45°, acute
4. 35°, acute
5. 180°, straight
6. 80°, acute
7. 60°, acute
8. 160°, obtuse
9. 95°, acute
10. 125°, obtuse

Problem Solving: 87 boxes

Geometry—Solutions

Page 20

Review Exercises
1. 205
2. 221
3. 1,155
4. 22,176
5. 2,079
6. 1,118

Answers for S1 - 10 are approximate.

S1. 43°, acute
S2. 104°, obtuse
1. 60°, acute
2. 74°, acute
3. 106°, obtuse
4. 120°, obtuse
5. 77°, acute
6. 137°, obtuse
7. 160°, obtuse
8. 36°, acute
9. 113°, obtuse
10. 20°, acute

Problem Solving: 12 gallons

Page 21

Review Exercises
1. answers vary
2. answers vary
3. answers vary
4. answers vary
5. 12°
6. 28°

Answers for S1 - 10 are approximate.

S1. 23°, acute
S2. 104°, obtuse
1. 60°, acute
2. 127°, obtuse
3. 77°, acute
4. 45°, acute
5. 118°, obtuse
6. 104°, obtuse
7. 77°, acute
8. 144°, obtuse
9. 54°, acute
10. 11°, acute

Problem Solving: 8 gallons, $24

Page 22

Review Exercises
1. 37°
2. 6,776
3. answers vary
4. answers vary
5. answers vary
6. answers vary

S1. 39°
S2. 93°
1. 18°
2. 74°
3. 85°
4. 38°
5. 19°
6. 105°
7. 168°
8. 165°
9. 8°
10. 67°

Problem Solving: 21 miles

Page 23

Review Exercises
1. 47°
2. 73°
3. 1,242
4. answers vary
5. 1,748
6. 3,969

S1. 115°
S2. 31°
1. 101°
2. 39°
3. 147°
4. 161°
5. 118°
6. 78°
7. 47°
8. 79°
9. 34°
10. 42°

Problem Solving: 8,000 sheets

Page 24

Review Exercises
1. 76°
2. 166°
3. 32°
4. 28°
5. 266
6. 6,755

S1. answers vary
S2. answers vary
1. answers vary
2. answers vary
3. answers vary
4. answers vary
5. answers vary
6. answers vary
7. answers vary

Problem Solving: $54,000

Page 25

Review Exercises
1. 98°
2. 73°
3. 5,505
4. 39°
5. 142°
6. answers vary

S1. answers vary
S2. answers vary
1. answers vary
2. answers vary
3. answers vary
4. answers vary
5. answers vary
6. answers vary
7. answers vary

Problem Solving: $46

Page 26

Review Exercises
1. ∠ DEF, acute
2. ∠ HGF, obtuse
3. ∠ JKL, right
4. parallel
5. perpendicular
6. 75°

S1. rectangle
S2. triangle
1. square, rectangle, parallelogram
2. rectangle, parallelogram
3. trapezoid
4. triangle
5. trapezoid
6. square, rectangle, parallelogram
7. parallelogram
8. rectangle, parallelogram
9. triangle
10. trapezoid

Problem Solving: 7 boxes, 6 left over

Page 27

Review Exercises
1. 76°
2. rectangle, square, parallelogram
3. 68°
4. answers vary
5. answers vary
6. answers vary

S1. square, rectangle, parallelogram
S2. rectangle, parallelogram
1. trapezoid
2. triangle
3. trapezoid
4. trapezoid
5. square, rectangle, parallelogram
6. rectangle, parallelogram
7. parallelogram
8. parallelogram
9. triangle
10. trapezoid

Problem Solving: 57 desks

Geometry—Solutions

Page 28

Review Exercises

1. triangle
2. trapezoid
3. square, rectangle, trapezoid, parallelogram
4. square, rectangle, parallelogram
5. 72°
6. 44°

S1. scalene, right
S2. equilateral, acute
1. scalene, obtuse
2. isosceles, acute
3. isosceles, right
4. scalene, acute
5. equilateral, acute
6. scalene, obtuse
7. scalene, right
8. isosceles, acute
9. equilateral, acute
10. isosceles, right

Problem Solving: 180°

Page 29

Review Exercises

1. obtuse
2. equilateral
3. scalene
4. isosceles
5. 60°
6. scalene

S1. equilateral, acute
S2. scalene, acute
1. isosceles, right
2. scalene, obtuse
3. equilateral, acute
4. isosceles, acute
5. scalene, right
6. isosceles, right
7. equilateral, acute
8. scalene, acute
9. isosceles, acute
10. scalene, right

Problem Solving: 41°

Page 30

Review Exercises

1. scalene
2. right
3. isosceles, acute
4. 4,536
5. 156
6. 166

S1. 34 ft.
S2. 29 ft.
1. 47 ft.
2. 48 ft.
3. 33 ft.
4. 54 ft.
5. 70 ft.
6. 41 cm.
7. 225 mi.
8. 34 ft.
9. 86 ft.
10. 34 ft.

Problem Solving: 54 ft.

Page 31

Review Exercises

1. obtuse
2. acute
3. right
4. acute
5. scalene
6. 63°

S1. 26 in.
S2. 38 cm.
1. 24 ft.
2. 22 cm.
3. 17 ft.
4. 44 in.
5. 90 cm.
6. 21 ft.
7. 168 cm
8. 27 ft.
9. 32 in.
10. 89 cm

Problem Solving: 102 ft.

Page 32

Review Exercises

1. 42 ft.
2. 76 in.
3. 60 ft.
4. 96 ft.
5. 40°
6. 117°

S1. 100 ft.
S2. 520 ft.
1. 384 ft.
2. 108 in.
3. 39 ft.
4. 171 in.
5. 23 ft.
6. $720
7. 97 in.

Problem Solving: $4.99

Page 33

Review Exercises

1. 128 ft.
2. 104 ft.
3. 85 ft.
4. 339 in.
5. 102 in.
6. 412 ft.

S1. 58 ft.
S2. 24 sections
1. 68 ft.
2. 16 in.
3. 8 hrs.
4. 12 ft.
5. 190 ft.
6. 126 in.
7. 48 in.

Problem Solving: 450 miles per hour

Page 34

Review Exercises

1. scalene
2. right
3. 167°
4. 108 ft.
5. 60°
6. 16 ft.

S1. diameter
S2. $\overline{VT}, \overline{YR}, \overline{XS}$
1. radius
2. chord
3. $\overline{CD}, \overline{DE}, \overline{DG}, \overline{DF}$
4. $\overline{AB}, \overline{GF}, \overline{CE}$
5. 8 ft.
6. P
7. $\overline{RY}, \overline{VT}, \overline{XS}$
8. 48 ft.
9. $\overline{PX}, \overline{PS}, \overline{PZ}$
10. $\overline{XS}$

Problem Solving: 13 mi.

Page 35

Review Exercises

1. 32 ft.
2. 9 ft.
3. 18.84
4. 66
5. 50°
6. scalene, right

S1. radius
S2. $\overline{AH}, \overline{AB}, \overline{AE}$
1. chord
2. chord
3. $\overline{GF}, \overline{HE}, \overline{CD}$
4. $\overline{XT}$
5. 36 ft.
6. 48 in.
7. $\overline{WX}, \overline{WT}, \overline{WQ}$
8. $\overline{RS}, \overline{XT}$
9. A
10. $\overline{AH}, \overline{AE}$

Problem Solving: $672

Geometry—Solutions

Page 36

Review Exercises
1. 24 ft.
2. 110
3. 25.12
4. 28 in.
5. scalene
6. isosceles

S1. 12.56 ft.
S2. 44 ft.
1. 18.84 ft.
2. 25.12 ft.
3. 28.26 ft.
4. 88 ft.
5. 37.68 ft.
6. 31.4 ft.
7. 12.56 ft

Problem Solving: $5,600

Page 37

Review Exercises
1. 292 in.
2. 100 ft.
3. 315 m.
4. 87 ft.
5. 368 ft.
6. 672 ft.

S1. 88 ft.
S2. 28.26 ft.
1. 47.1 ft.
2. 50.24 ft.
3. 66 ft.
4. 132 ft.
5. 110 ft.
6. 37.68 ft.
7. 314 ft.

Problem Solving: 88 ft.

Page 38

Review Exercises
1. 88 ft.
2. 90 ft.
3. 44 ft.
4. 8
5. 21
6. obtuse

S1. 110 ft.
S2. 44 in.
1. 44 in.
2. 88 mi.
3. 22 ft.
4. 4 ft.
5. 3 ft.
6. 22 ft.
7. 31.4 ft.

Problem Solving: 124 ft.

Page 39

Review Exercises
1. 67°
2. 83°
3. trapezoid
4. 28
5. 12
6. 71°

S1. 35 ft.
S2. 15 ft.
1. 28 ft.
2. 352 m. or 351.68 m.
3. 25,120 mi.
4. 31.4 ft.
5. 176 ft.
6. 84 ft.
7. 20 ft.

Problem Solving: 240 ft.

Page 40

Review Exercises
1. 170 ft.
2. 18.84 ft.
3. 192 ft.
4. 88 ft.
5. 51 ft.
6. 76 ft.

S1. 54 ft.
S2. 88 ft.
1. 66 ft.
2. 28 ft.
3. 116 ft.
4. 21 ft.
5. 51 ft.
6. 25.12 ft.
7. 177 ft.
8. 48 ft.
9. 460 ft.
10. 40 ft.

Problem Solving: 7 hrs.

Page 41

Review Exercises
1. acute
2. obtuse
3. right
4. 8
5. pentagon
6. decagon

S1. 116 ft.
S2. 37.68
1. 110 ft.
2. 258 ft.
3. 88 ft.
4. 185 ft.
5. 460 ft.
6. 390 ft.
7. 284 ft.
8. 132 ft.
9. 252 ft.
10. 132 ft.

Problem Solving: $306

Page 42

Review Exercises
1. answers vary
2. answers vary
3. 75°
4. 165°
5. 45°
6. 127°

S1. 110 ft.
S2. 85 ft.
1. 210 ft.
2. 49 ft.
3. 720 ft.
4. 38 ft.
5. $1,400
6. 156 ft.
7. 184 ft.

Problem Solving:
29,050 gallons

Page 43

Review Exercises
1. 64 ft.
2. 90 ft.
3. 18.84 ft.
4. 66 ft.
5. 140 ft.
6. 99 ft.

S1. 63 ft.
S2. 3 ft.
1. 125 laps
2. 4,605 ft.
3. 192 in.
4. 12 ft.
5. 4 laps
6. 304 ft.
7. 66 ft.

Problem Solving: 160 mi.

Geometry—Solutions

Page 44

Review Exercises
1. answers vary
2. acute, obtuse, right, straight
3. radius, diameter, chord, center
4. scalene, isosceles, equilateral
5. acute, obtuse, right
6. pentagon, hexagon, octagon, decagon

S1. 5 ft.
S2. 9 cm.
1. 7 ft.
2. 10 ft.
3. 8 ft.
4. 19 ft.
5. 9 ft.
6. 12 ft.
7. 16 ft.

Problem Solving: $14.15

Page 45

Review Exercises
1. 68 ft.
2. 52 ft.
3. 174 ft.
4. 37.68 ft.
5. 22 ft.
6. 525 ft.

S1. 8 ft.
S2. 13 ft.
1. 9 ft.
2. 8 ft.
3. 18 cm.
4. 28 cm.
5. 13 ft.
6. 35 cm.
7. 20 ft.

Problem Solving: 230 cartons

Page 46

Review Exercises
1. 68 ft.
2. 476 ft.
3. 132 ft.
4. 25.12 ft.
5. 6
6. 8

S1. 8 ft.^{2}
S2. 19 ft.^{2}
1. 16 ft.^{2}
2. 18 ft.^{2}
3. 27 ft.^{2}
4. 21 ft.^{2}
5. 20 ft.^{2}
6. 24 ft.^{2}
7. 13 ft.^{2}

Problem Solving: 92

Page 47

Review Exercises
1. equilateral
2. obtuse
3. 9 ft.
4. 28.26 ft.
5. 132 ft.
6. 288 ft.

S1. answers vary
S2. answers vary
1. answers vary
2. answers vary
3. answers vary
4. answers vary
5. answers vary
6. answers vary
7. answers vary

Problem Solving: $21,200

Page 48

Review Exercises
1. 84°
2. 163°
3. square, rectangle, parallelogram
4. 145 ft.
5. 64 ft.
6. 12.56 ft.

S1. 169 sq. ft.
S2. 165 sq. ft.
1. 84 sq. ft.
2. 400 sq. ft.
3. 132 sq. ft.
4. 391 sq. ft.
5. 13.5 sq. ft.
6. 625 sq. ft.
7. 32,400 sq. ft.

Problem Solving: 842

Page 49

Review Exercises
1. 143 sq. ft.
2. 225 sq. ft.
3. 62 ft.
4. 36 ft.
5. 88 ft.
6. 31.4 ft.

S1. 209 ft.^{2}
S2. 256 ft.^{2}
1. 126 ft.^{2}
2. 625 ft.^{2}
3. 475 ft.^{2}
4. 400 ft.^{2}
5. 14 ft.
6. 25 ft.
7. 30,000 cm.^{2}

Problem Solving: 285 sq. ft.

Page 50

Review Exercises
1. 68 ft.
2. 289 ft.^{2}
3. 160 ft.^{2}
4. 52 ft.
5. 11 ft.
6. 50°

S1. 78 ft.^{2}
S2. 77 ft.^{2}
1. 54 ft.^{2}
2. 176 ft.^{2}
3. 17.5 ft.^{2}
4. 91 ft.^{2}
5. 84 ft.^{2}
6. 117 ft.^{2}
7. 71.5 ft.^{2}

Problem Solving: $66,000

Page 51

Review Exercises
1. 60 ft.
2. 76 ft.
3. 66 ft.^{2}
4. 42 ft.^{2}
5. 126 ft.^{2}
6. 37.68 ft.

S1. 17.5 ft.^{2}
S2. 65 ft.^{2}
1. 64 ft.^{2}
2. 325 ft.^{2}
3. 71.5 ft.^{2}
4. 264 ft.^{2}
5. 36 ft.^{2}
6. 120 cm.^{2}
7. 16 ft.

Problem Solving: 320 ft.^{2}

Geometry—Solutions

Page 52

Review Exercises
1. 196 ft.^{2}
2. 96 ft.^{2}
3. 54 ft.^{2}
4. 105 ft.^{2}
5. 52.5 ft.^{2}
6. 64 ft.^{2}

S1. 25 ft.^{2}
S2. 40 ft.^{2}
1. 36 ft.^{2}
2. 44 ft.^{2}
3. 70 ft.^{2}
4. 144 ft.^{2}
5. 80 ft.^{2}
6. 16.5 ft.^{2}
7. 330 ft.^{2}

Problem Solving: 240 in.^{2}

Page 53

Review Exercises
1. 32 ft.
2. 46 ft.
3. 85 ft.
4. 88 ft.
5. 230 ft.
6. 132 ft.

S1. 49 ft.^{2}
S2. 30 ft.^{2}
1. 85 ft.^{2}
2. 800 cm.^{2}
3. 1,650 ft.^{2}
4. 64 ft.^{2}
5. 26 ft.^{2}
6. 9 ft.^{2}
7. 42 cm.^{2}

Problem Solving: 2,505 pounds

Page 54

Review Exercises
1. 54 ft.^{2}
2. 66 cm.^{2}
3. 38 ft.
4. 56 ft.
5. 196 ft.^{2}
6. 84 ft.^{2}

S1. 50.24 ft.^{2}
S2. 113.04 ft.^{2}
1. 78.5 ft.^{2}
2. 616 ft.^{2}
3. 12.56 ft.^{2}
4. 200.96 ft.^{2}
5. 78.5 ft.^{2}
6. 113.04 ft.^{2}
7. 154 ft.^{2}

Problem Solving: 320 ft.

Page 55

Review Exercises
1. obtuse
2. acute
3. 110°
4. equilateral, acute
5. 18.84 ft.
6. 28.26 ft.^{2}

S1. 1,386 ft.^{2}
S2. 28.26 ft.^{2}
1. 314 ft.^{2}
2. 1,256 ft.^{2}
3. 616 ft.^{2}
4. 3.14 ft.^{2}
5. 254.34 ft.^{2}
6. 1,386 ft.^{2}
7. 379.94 ft.^{2}

Problem Solving: 20 classes

Page 56

Review Exercises
1. 154 ft.^{2}
2. 196 ft.^{2}
3. 135 ft.^{2}
4. 31.5 ft.^{2}
5. 96 ft.^{2}
6. 72 ft.^{2}

S1. P = 38 ft., A = 84 ft.^{2}
S2. P = 27 ft., A = 25 ft.^{2}
1. P = 48 ft., A = 144 ft.^{2}
2. P = 44 ft., A = 120 ft.^{2}
3. P = 38 ft., A = 72 ft.^{2}
4. C = 18.84 ft., A = 28.26 ft.^{2}
5. C = 44 ft., A = 154 ft.^{2}
6. P = 24 ft., A = 24 ft.^{2}
7. P = 27 ft., A = 37.5 ft.^{2}

Problem Solving: $396

Page 57

Review Exercises
1. 60 ft.
2. 84 ft.
3. 25.12 ft.
4. 36 ft.
5. 44 ft.
6. 51 ft.

S1. P = 23 ft., A = 30 ft.^{2}
S2. C = 88 ft., A = 616 ft.^{2}
1. P = 32 ft., A = 54 ft.^{2}
2. P = 27 ft., A = 35 ft.^{2}
3. P = 48 ft., A = 144 ft.^{2}
4. C = 25.12 ft., A = 50.24 ft.^{2}
5. P = 86 ft., A = 406 ft.^{2}
6. P = 47 ft., A = 112.5 ft.^{2}
7. P = 28 ft., A = 42 ft.^{2}

Problem Solving: $1.35

Page 58

Review Exercises
1. 256 ft.^{2}
2. 72 ft.^{2}
3. 352 ft.^{2}
4. 126 ft.^{2}
5. 288 ft.^{2}
6. 78.5 ft.^{2}

S1. 52 ft.
S2. 20 bags
1. 96 ft.
2. 14 sections
3. 50.24 ft.
4. 3 cans
5. 156 ft.
6. 16 mi.
7. 630 meters

Problem Solving: $1,200

Page 59

Review Exercises
1. 55°
2. 102°
3. 36 ft.
4. 235 ft.
5. 24 ft.
6. P = 138 ft., A = 884 ft.^{2}

S1. 6.28 mi.
S2. $1,440
1. 132 mi.
2. 40,000 sq. ft.
3. 96 in.
4. 28.26 in.^{2}
5. 144 in.
6. 56 ft.^{2}
7. 144 in.^{2}

Problem Solving: 110 minutes

Geometry—Solutions

Page 60

Review Exercises
1. 25.12 ft.
2. 112 ft.^{2}
3. 380 cm.
4. 48 ft.^{2}
5. 66 ft.
6. 82.5 ft.^{2}

S1. rectangular prism, 6, 12, 8
S2. square pyramid, 5, 8, 5
1. cone, 1, 1, 1
2. cube, 6, 12, 8
3. cylinder, 2, 2, 0
4. triangular pyramid, 4, 6, 4
5. triangular prism, 5, 8, 6
6. sphere
7. 1

Problem Solving: $7.68

Page 61

Review Exercises
1. square pyramid, triangular prism
2. cube, rectangular prism
3. sphere
4. cube, rectangular prism
5. square pyramid
6. cone, cylinder

S1. answers vary
S2. answers vary
1. answers vary
2. answers vary
3. answers vary
4. answers vary
5. answers vary
6. answers vary
7. answers vary
8. answers vary
9. answers vary
10. answers vary

Problem Solving: 180 in.

Page 62

Review Exercises
1. 72°
2. 66°
3. 78.5 ft.^{2}
4. 31.5 ft.^{2}
5. 90 ft.
6. 256 ft.^{2}

S1. 24 ft.^{2}
S2. 136 ft.^{2}
1. 150 m.^{2}
2. 176 ft.^{2}
3. 294 ft.^{2}
4. 356 cm.^{2}
5. 258 cm.^{2}
6. 30 m.^{2}
7. 280 in.^{2}

Problem Solving: 84 mi.^{2}

Page 63

Review Exercises
1. 50.24 ft.^{2}
2. 28.26 ft.^{2}
3. 441 ft.^{2}
4. 540 cm.^{2}
5. 330 ft.^{2}
6. 104 ft.^{2}

S1. 34 ft.^{2}
S2. 96 m.^{2}
1. 40 ft.^{2}
2. 216 ft.^{2}
3. 270 m.^{2}
4. 66 ft.^{2}
5. 90 in.^{2}
6. 166 cm.^{2}
7. 62 in.^{2}

Problem Solving: $19,200

Page 64

Review Exercises
1. 116 ft.
2. 46 ft.
3. 44 ft.
4. 18.84 ft.
5. 70 ft.
6. 114 ft.

S1. 336 cm.^{3}
S2. 64 ft.^{3}
1. 240 cm.^{3}
2. 343 in.^{3}
3. 4,480 in.^{3}
4. 80 ft.^{3}
5. 273 cm.^{3}
6. 1,680 ft.^{3}
7. 2,808 ft.^{3}

Problem Solving: 4,550 mi.

Page 65

Review Exercises
1. 142 ft.^{2}
2. 192 ft.^{3}
3. 52°
4. 31°
5. 27°
6. 31.4 cm.

S1. 360 in.^{3}
S2. 343 ft.^{3}
1. 560 ft.^{3}
2. 729 in.^{3}
3. 1,200 ft.^{3}
4. 1,331 cm.^{3}
5. 4,096 in.^{3}
6. 2,200 in.^{3}
7. 280 ft.^{3}

Problem Solving: $117,000

Page 66

Review Exercises
1. 242 ft.^{2}
2. 84 in.^{3}
3. 96 ft.^{2}
4. 64 ft.^{3}
5. 228 cm.
6. 361 ft.^{2}

S1. 125 m.^{3}
S2. 150 m.^{2}
1. 258 cm.^{2}
2. 270 cm.^{3}
3. 8 ft.^{3}
4. 24 ft.^{2}
5. 126 ft.^{2}
6. 90 ft.^{3}
7. 216 ft.^{3}

Problem Solving: 6 days

Page 67

Review Exercises
1. 72 ft.^{2}
2. 342 cm.^{2}
3. 256 in.^{2}
4. 135 in.^{2}
5. 40.5 ft.^{2}
6. 50.24 ft.^{2}

S1. 756 ft.^{3}
S2. 96 ft.^{2}
1. 412 ft.^{2}
2. 486 in.^{2}
3. 1,500 in.^{3}
4. 148 in.^{2}
5. 13 in.

Problem Solving: $33,120

Geometry—Solutions

Page 68

Review Exercises

1. answers vary
2. answers vary
3. answers vary
4. answers vary
5. answers vary
6. triangle FIK
7. answers vary
8. answers vary
9. answers vary
10. trapezoid LKIJ
11. equilateral, acute
12. scalene, obtuse
13. $\overline{CG}$, $\overline{AF}$
14. $\overline{EC}$, $\overline{EF}$, $\overline{EA}$, $\overline{EG}$
15. $\overline{BH}$, $\overline{CG}$, $\overline{AF}$
16. 12 ft.
17. chord
18. rectangular prism, 6, 12, 8
19. triangular prism, 5, 9, 6
20. cube, 6, 12, 8

Page 69

Review Exercises

1. answers vary
2. answers vary
3. answers vary
4. answers vary
5. $\triangle$BCD, $\triangle$FGH, $\triangle$ABF
6. trapezoid GHJI
7. answers vary
8. pentagon BDEHF
9. answers vary
10. answers vary
11. scalene, acute
12. scalene, right
13. hypotenuse
14. trapezoid
15. square, rectangle, parallelogram
16. octagon
17. regular
18. square pyramid, 5, 8, 5
19. cylinder 2, 2, 0
20. cone, 1, 1, 1

Page 70

Review Exercises

1. 52 ft.
2. 384 ft.
3. 44 ft.
4. 174 ft.
5. 114 ft.
6. 37.68 ft.
7. 240 ft.^{2}
8. 56 ft.^{2}
9. 192 ft.^{2}
10. 169 ft.^{2}
11. 45 ft.^{2}
12. 154 ft.^{2}
13. 50.24 ft.^{2}
14. 49.5 ft.^{2}
15. 48 ft.^{2}
16. 289 ft.^{2}
17. 216 cm.^{2}
18. 152 ft.
19. 236 cm.^{2}
20. 750 cm.^{3}

Page 71

Review Exercises

1. 53 ft.
2. 88 ft.
3. 166 ft.
4. 34.54
5. 260 ft.
6. 290 ft.
7. 32.5 ft.^{2}
8. 154 ft.^{2}
9. 28.26 ft.^{2}
10. 119 ft.^{2}
11. 136 ft.^{2}
12. 27.5 ft.^{2}
13. 27 ft.^{2}
14. 256 ft.^{2}
15. 17.5 ft^2
16. 25 ft.^{2}
17. 100 ft.^{2}
18. 720 ft.^{2}
19. 112 ft.^{2}
20. 216 ft.^{3}

Problem Solving—Solutions

Page 74

Review Exercises
1. 605
2. 684
3. 1,130
4. 1,944
5. 30,018
6. 44,674

S1. April
S2. 10°
1. July
2. July
3. June
4. 7° - 8°
5. August
6. March; April
7. May; July
8. 13°
9. 3° - 4°
10. March; April; July

Page 75

Review Exercises
1. 1,520
2. 427
3. 4,494
4. 1,075
5. 940
6. 2,508

S1. Auberry; Winston
S2. 1,050
1. 1,200
2. 50
3. 350
4. 3,050 - 3,075
5. Approx. 325
6. 450
7. Approx. 1,375
8. 50
9. Sun City
10. 1,700

Page 76

Review Exercises
1. 15,630
2. 1,472
3. 9,541
4. 3,772
5. 205 r2
6. 47 r1

S1. 90
S2. 10
1. Tests 4, 6, and 7
2. 20
3. Approx. 91
4. 4
5. 80, 85
6. 90
7. 5
8. 15
9. 3
10. improved

Page 77

Review Exercises
1. 9,872
2. 4,278
3. 11,310
4. 201
5. 399 r3
6. 180 r5

S1. July; September
S2. 100
1. 700
2. 100
3. May
4. June
5. 100
6. July; September
7. 1,100
8. 200
9. 100
10. April; May

Page 78

Review Exercises
1. 15,272
2. 4,144
3. 613
4. 3 r6
5. 15 r 15
6. 201

S1. 23%
S2. 70%
1. 33%
2. $460
3. $200
4. $400
5. 68%
6. 32%
7. $24,000
8. 80%
9. car, clothing
10. other expenses

Page 79

Review Exercises
1. 38
2. 145
3. 11 r25
4. 9 r67
5. 117 r6
6. 205 r2

S1. 4/24 = 1/6
S2. 6/24 = 1/4
1. 5
2. 10
3. 8
4. 9/24 = 3/8
5. 16/24 = 2/3
6. 90
7. 6:00 A.M.
8. 2:30 P.M.
9. 40
10. 6/24 = 1/4

Page 80

Review Exercises
1. 1,074
2. 5,644
3. 89,790
4. 21 r1
5. 21
6. 21 r1

S1. 6,000
S2. 4,500
1. 1989
2. 13,500
3. 1989; 1991
4. 7,000
5. 11,000
6. $150,000
7. $600,000
8. 5,000
9. 19,500
10. 3,000

Page 81

Review Exercises
1. 17 r1
2. 241 r6
3. 23 r7
4. 71 r27
5. 349 r17
6. 225 r11

S1. 1990
S2. approx. 6-7 hours
1. 45
2. approx 5
3. 2,000
4. 1960
5. 1950; 1960
6. approx. 2
7. 8
8. approx. 12
9. approx. 5
10. approx 10

Problem Solving—Solutions

Page 82

Review Exercises
1. Ken
2. 70
3. 40
4. 200
5. Sue
6. 30%
7. 25%
8. 55%
9. Food
10. 15%
11. 60
12. cat
13. fish
14. 30
15. dog
16. 400
17. August
18. 400
19. 500
20. July; August

Page 83

Review Exercises
1. Grade 4
2. Approx. 1,500 cans
3. Approx. 9,500 cans
4. Approx. 21,500 cans
5. Grade 2
6. 20 students
7. 11 more C's
8. 100 students
9. 13 more C's and B's
10. B's
11. 25 inches
12. Approx. 9 inches
13. January
14. 60 inches
15. December
16. 500 boxes
17. troop 15
18. 200 boxes
19. 900 boxes
20. 750 boxes

Page 84

Review Exercises
S1. 29,710
S2. 3,655
1. 547
2. 573
3. 10,103
4. 7,243,008
5. 5,665
6. 1,196
7. 3,418
8. 1,845
9. 616
10. 90,558

Page 85

Review Exercises
S1. 29,025
S2. 220 r9
1. 228
2. 30,612
3. 22,572
4. 232,858
5. 141 r2
6. 282 r5
7. 25 r19
8. 94 r2
9. 214 r4
10. 56 r9

Page 86

Review Exercises
1. 1,236
2. 7,163
3. 2,127
4. 391
5. 58 r1
6. 195 r8

S1. 138
S2. 1,580
1. 391 increase
2. $139
3. 14,532 feet
4. 55 mph
5. 682 miles

Page 87

Review Exercises
1. 9,152
2. 207,936
3. 424
4. 13 r15
5. 118 r43
6. 142 r23

S1. 173 pieces
S2. 5 buses
1. $27
2. 21,500 gallons
3. 125 pounds
4. 1,650 cars
5. 450 mph

Page 88

Review Exercises
1. 1 r 14
2. 17 r22
3. 237 r22
4. 2 r1
5. 32 r5
6. 315 r1

S1. $705
S2. $312
1. 416 votes
2. 866 feet
3. $27
4. 197 containers
5. 840 students

Page 89

Review Exercises
1. 632
2. 5,705
3. 305
4. 124
5. 2,982
6. 6,754

S1. 1,032
S2. $640
1. 10,940
2. 180 cows
3. 1,219 students
4. 24 students
5. 89

Problem Solving—Solutions

Page 90

Review Exercises
1. 55 mph
2. 4,368
3. 750 students
4. 1,259
5. $1,061
6. 6,437

S1. 90
S2. 1,985 pounds
1. $20,280
2. 117 hours
3. 356 seats
4. 186,650 gallons
5. $42

Page 91

Review Exercises
1. 14,766
2. 86,892 books
3. 23 r15
4. 5,460 miles
5. 164 r8
6. $6,695

S1. $385
S2. 8 boxes, 8
1. 72 seats
2. 4 pieces
3. $97
4. 24 classes
5. 2,064 hours

Page 92

Review Exercises
1. 18 boxes
2. 6,199
3. 116 feet
4. 7,280
5. 376, 94
6. 339 r2

S1. $8
S2. 17 miles
1. 470 feet
2. 720 bushels
3. $1,200
4. 4 buses
5. 87

Page 93

Review Exercises
1. 21
2. 50 students
3. 1,261
4. $13
5. 42,350
6. 95

S1. $148
S2. $250
1. $2,720
2. $15,850
3. 1,485 miles
4. 51,925 gallons
5. $89,500

Page 94

Review Exercises
1. 80
2. 11,421
3. 384 students
4. 201
5. 1,920
6. 8,722

S1. $39
S2. 80 feet; $560
1. 480 miles
2. 17 hrs. 10 min.
3. $600
4. $900
5. $163

Page 95

Review Exercises
1. 9,732
2. 550 mph
3. 85
4. 2,180
5. $36
6. $1,618
7. 18 classes
8. 30 payments
9. 272 feet; $850
10. $2,520

Page 96

Review Exercises
S1. 1 3/7
S2. 10 1/4
1. 7/18
2. 13/18
3. 4 2/5
4. 8 5/8
5. 11 1/4
6. 4 3/4
7. 8 7/15
8. 3 1/2
9. 11/16
10. 8 7/12

Page 97

Review Exercises
S1. 3
S2. 3 2/3
1. 3/26
2. 27
3. 1 7/8
4. 8 1/6
5. 1 1/2
6. 7
7. 2 4/9
8. 3 1/3
9. 2 4/7
10. 1 1/3

Page 98

Review Exercises
1. 5/6
2. 3 13/15
3. 5 1/6
4. 2 1/6
5. 5 2/15
6. 5 4/5

S1. 5 3/4 cups
S2. $40
1. 8 1/2 pounds
2. 11 pieces
3. 30 miles
4. 29 1/4 minutes
5. 34 feet

Page 99

Review Exercises
1. 2/5
2. 5/6
3. 3
4. 1 1/3
5. 9
6. 1 3/4

S1. 15 pounds
S2. 1 1/2 pounds
1. 14 1/4 hours
2. 125 miles
3. 16 tires
4. 27 pounds
5. 4 3/4 hours

Page 100

Review Exercises
1. 1 1/10
2. 5/8
3. 6 1/6
4. 6 2/15
5. 5 1/10
6. 3 7/10

S1. $78
S2. 12 packages
1. 1 1/4 hours
2. 63 1/4 inches
3. 35 1/6 pages
4. 13 1/3 pounds
5. 90 miles

Page 101

Review Exercises
1. 4/11
2. 11
3. 2
4. 1 1/2
5. 1 1/3
6. 2

S1. 33 feet
S2. 30 pounds
1. 5 1/5 feet
2. 1/2 pound
3. 3 5/6 miles
4. 27 1/2 feet
5. $2,100

Problem Solving—Solutions

Page 102

Review Exercises
1. 1 3/4 inches
2. 1 1/12
3. 21 pages
4. 5/8
5. 6 1/4 pounds
6. 5 1/4

S1. 4 yards
S2. $14
1. 1 1/4 gallons
2. 18 girls
3. $147
4. 24 bracelets
5. 2,000 acres

Page 103

Review Exercises
1. 14
2. 11 patties
3. 3 1/3
4. 10 yards
5. 3
6. 23 1/12 minutes

S1. 60 miles
S2. $48
1. 17 bushels
2. 1 1/2 pounds
3. $120,000
4. $2,600
5. 112 pages

Page 104

Review Exercises
1. 1 7/8
2. $48
3. 1,299
4. 520 miles
5. 21 r9
6. 160 11/12 pounds

S1. 1/6
S2. 406 pages
1. 5 1/2 miles
2. 40, $80
3. 1 1/2 pounds
4. 27 3/4 hours
5. $20

Page 105

Review Exercises
1. 6 1/4
2. 8/15
3. 1 11/16
4. 8 3/4
5. 2 1/4
6. 1 7/8

S1. $19
S2. 9 3/4 dollars
1. $36
2. $51
3. 6 pounds
4. $8
5. $840

Page 106

Review Exercises
1. 15
2. 28 r16
3. 7 1/2
4. 11,102
5. 7/30
6. 5,409

S1. 23 students
S2. 352 lots, $1,760,000
1. 510 students
2. 200 acres
3. $800
4. $240
5. $60,000

Page 107

Review Exercises
1. $48
2. 116 3/4 pounds
3. 8 pieces
4. 10 5/12 hours
5. $147
6. $47
7. 20 packages, $100
8. 150; 100
9. 1/6 pie
10. $725

Page 108

Review Exercises
S1. 18.38
S2. 3.687
1. 23.7114
2. 6.17
3. 16.3
4. 8.24
5. 2.08
6. 19.56
7. 2.2
8. 4.487
9. 10.396
10. 25.7

Page 109

Review Exercises
S1. 7.452
S2. .65
1. 12.78
2. 54.4
3. 4.928
4. 147.888
5. 1.34
6. 1.46
7. 400
8. .75
9. .05
10. .013

Page 110

Review Exercises
1. 11.243
2. 20.95
3. 4.478
4. 2.636
5. 6.42
6. 18.56

S1. $63.60
S2. 500 mph
1. $.34
2. $518.85
3. 194 feet
4. $1.14
5. $12.60

Page 111

Review Exercises
1. .216
2. .0786
3. .000006
4. 1.84
5. .026
6. .007

S1. 360 miles
S2. 7 pounds
1. 278.35 pounds
2. $.89
3. 11.2 gallons
4. 20 cans
5. $3.36

Page 112

Review Exercises
1. 6.7
2. 6.5
3. 170
4. 5.4
5. 9.2
6. 3.2

S1. $15.84
S2. $47.70
1. $2.25
2. $1931.24
3. $75.48
4. $250.50
5. $9.84

Page 113

Review Exercises
1. 13.24
2. 3.287
3. 2.562
4. $22
5. $3.15
6. 250

S1. $620
S2. 41.6 inches
1. $723.59
2. $35.15
3. 13.53 inches
4. $3.20
5. $3.40

Problem Solving—Solutions

Page 114

Review Exercises
1. .6
2. $30
3. 14/15
4. 2 1/2 feet
5. 1 9/14
6. 30 inches

S1. $13.40
S2. $8.46
1. $24.80
2. $625.75
3. $7.14
4. $5.10
5. $7.05

Page 115

Review Exercises
1. 1825 miles
2. 1 1/7
3. 408 engines
4. 9 1/3
5. 2 1/2 feet
6. 2

S1. $12.84
S2. $37.79
1. $2.62
2. 12.1
3. $18
4. $8.95
5. $468

Page 116

Review Exercises
1. 53 mph
2. 2,656
3. 3,504
4. 7,982
5. 15 miles
6. 1,193

S1. 25.02
S2. 19.32
1. 5.31
2. 16.884
3. $504
4. $14.95
5. $1.25

Page 117

Review Exercises
1. 11.22
2. 4.835
3. 21.288
4. .26
5. .05
6. 5,000

S1. 7.9 miles
S2. $5,600,000
1. $45
2. $4.70
3. $147
4. $170
5. $44.79

Page 118

Review Exercises
1. 5 1/2
2. 1 3/8
3. 3 7/12
4. 9 3/40
5. 4 3/4
6. 4 3/4

S1. $38.79
S2. $960
1. 38 students
2. $79.92
3. 480 acres
4. 161.5 miles
5. $1.28

Page 119

1. 448 miles
2. $1.19
3. 7.8 pounds
4. $388.94
5. 9 pounds
6. $44.10
7. $19.34
8. $13.80
9. $10,500,000
10. $7.65

Page 120

S1. .2; 1/50
S2. .09; 9/100
1. .16; 4/25
2. .06; 3/50
3. .75; 3/4
4. .4; 2/5
5. .01; 1/100
6. .45; 9/20
7. .12; 3/25
8. .05; 1/20
9. .5; 1/2
10. .13; 13/100

Page 121

S1. 20%
S2. 20
1. 42
2. 21
3. 25%
4. 25
5. 20
6. 60
7. 75%
8. 16
9. 60
10. 20%

Page 122

Review Exercises
1. 433.8
2. 2.382
3. 7.2
4. 17.5
5. 15
6. 5

S1. 32 problems
S2. 470 students
1. $1,800
2. 9 stamps
3. $30,000
4. $4.20; $64.20
5. $12,000

Page 123

Review Exercises
1. 4.8
2. 48
3. 60%
4. 60%
5. 75%
6. 90%

S1. 1,500 acres
S2. $297.50
1. 180 girls
2. $17.28
3. 36 pages
4. $6.00
5. 32 games

Page 124

Review Exercises
1. 40%
2. 75%
3. 25%
4. 90%
5. 24
6. 9.6

S1. 80%
S2. 40%
1. 75%
2. 84%
3. 75%
4. 25%
5. 60%

Page 125

Review Exercises
1. 5.4
2. 80%
3. 15
4. 26
5. 30%
6. 24

S1. 75%
S2. 40%
1. 80%
2. 30%
3. 75%
4. 40%
5. 60%

Problem Solving—Solutions

Page 126

Review Exercises
1. 15
2. 160
3. 12
4. 120
5. 20%
6. 90%

S1. 25 games
S2. 120 stamps; 96 stamps
1. $30
2. 35
3. 1,000 bushels
4. 12 shots; 3 shots
5. 30 problems

Page 127

Review Exercises
1. 20
2. 40%
3. 225
4. 25
5. 80%
6. 22.5

S1. 50 questions
S2. 125 people
1. 20 marbles; 12 marbles
2. $25,000
3. 15
4. 50 students
5. 4 games

Page 128

1. 55
2. 30
3. 20%
4. 95%
5. 8 people
6. 75%
7. 35 students
8. 75%
9. 2,400 acres
10. 80%

Page 129

1. 60%
2. 40
3. 25%
4. 67.5
5. 25 games
6. 180 students
7. 75%
8. 56 gold fish
9. 75%
10. $34.56

Page 130

1. 814
2. 2,178
3. 23,965
4. 9,314
5. 7,693
6. 279
7. 3,628
8. 1,414
9. 1,714
10. 5,284
11. 292
12. 28,544
13. 1,665
14. 15,732
15. 158,178
16. 131 r2
17. 344
18. 14 r8
19. 253 r24
20. 69 r1

Page 131

1. 5/7
2. 1 1/4
3. 14/15
4. 5 7/8
5. 14 3/10
6. 3/4
7. 3 1/2
8. 2 6/7
9. 5 1/10
10. 4 11/12
11. 6/35
12. 3/44
13. 20
14. 2
15. 8 3/4
16. 2 1/2
17. 4 2/3
18. 1 1/3
19. 3 2/3
20. 4 1/2

Page 132

1. 11.913
2. 14.52
3. 30.7
4. 18.6
5. 2.61
6. 70.32
7. 7.92
8. 189.8
9. 1.512
10. .53298
11. 36.5
12. 3,600
13. 2.32
14. 1.34
15. .31
16. 300
17. .03
18. 13
19. .6
20. .35

Page 133

1. Grand Falls
2. 525 - 550 feet
3. 100 feet
4. Snake Falls
5. Morton Falls
6. 13%
7. 20%
8. $600
9. $300
10. 41%
11. 71° - 72°
12. 30°
13. June
14. August
15. 10°
16. 60,000
17. 40,000
18. 200,000
19. 180,000 pounds
20. salmon, cod, and snapper

Problem Solving—Solutions

Page 134

1. $3,514
2. 603
3. 50 problems
4. 12 bags
5. 3 3/4 dollars
6. 208 miles
7. $20.25
8. 103 pounds
9. $342.75
10. $2.81

Page 135

11. 28 correct
12. $30
13. $21.65
14. $3.80
15. 192 boys
16. $2,260
17. 25%
18. 4/15 pie
19. 15 students
20. 112 feet; $980

Page 136

1. 1,011
2. 2,919
3. 80,653
4. 10,433
5. 8,447
6. 376
7. 3,915
8. 4,415
9. 2,322
10. 6,323
11. 288
12. 30,612
13. 2,438
14. 22,572
15. 232,858
16. 141 r2
17. 282 r5
18. 25 r19
19. 214 r4
20. 56 r9

Page 137

1. 4/5
2. 1 1/3
3. 13/15
4. 8 2/9
5. 10 1/8
6. 1/2
7. 4 4/5
8. 4 2/5
9. 6 1/4
10. 5 14/15
11. 8/21
12. 3/26
13. 27
14. 1 7/8
15. 8 1/6
16. 1 1/2
17. 7
18. 2 4/9
19. 3 1/3
20. 2 4/7

Page 138

1. 12.283
2. 10.36
3. 29.1
4. 20.6
5. 4.73
6. 4.57
7. 9.36
8. 54.4
9. .752
10. 1.39956
11. 236
12. 2,700
13. 1.34
14. 1.46
15. .65
16. 400
17. .05
18. .013
19. .875
20. .44

Page 139

1. Joe
2. Amir, Jim
3. Joe
4. 9' 10"
5. 8 inches
6. 35%
7. 25%
8. $350
9. $50
10. 40%
11. 8
12. 4-5
13. February
14. 36
15. May
16. Friday
17. Tuesday
18. Tuesday
19. 160-170 Appr.
20. Wednesday, Friday

Page 140

1. 1,502 feet
2. 21 girls
3. 4,365 miles
4. 6 pieces
5. $1.38
6. $16.20
7. 86
8. 30 miles
9. 6 1/5 yards
10. 18 problems

Page 141

11. $9.11
12. $18.24
13. $39
14. 40 payments
15. $600
16. 20%
17. 410 students
18. $29.16
19. $2.91
20. 95%

Page 144

Review Exercises
1. 1,159
2. 436
3. 2,282

S1. No, members are countable
S2. No, members are uncountable
1. No, 3 is a member of each
2. Yes, can be paired 1-1
3. Answers vary
4. Answers vary
5. {3,5,7,9,11}
6. {0,2,4,6,8,10,12}
7. {3,4,5,6,7,8,9}
8. {10,15,20,25,30}
9. {1,3,5}
10. {8,9,10,11,12}

Problem Solving: 19 girls

Page 145

Review Exercises
1. Answers vary
2. Answers vary
3. Answers vary
4. Answers vary
5. No, 10 is a member of both sides
6. No, cannot be paired in a 1-1 correspondence

S1. Yes, all members of A are members of B
S2. {5,6,7}
1. {1,2,3,4,5,6,7,8}
2. No, not all members of A are members of B
3. {5}, {6}, {7}, {5,6}, {5,7}, {6,7}, {5,6,7}
4. {1,2,4,5,7}
5. {1,2,4,8}
6. {1,2,3,4,5,6,7}
7. {1,2,3,4,5,6,7,8,10}
8. {1,2,4,6}
9. Yes, can be paired in a 1-1 correspondence
10. No, 6 is common to both sets

Problem Solving: 80 cards

Page 146

Review Exercises
1. {2}
2. {1,2,3,4,6,8}
3. {1,2,3,6}
4. {2}
5. No, cannot be paired in a 1-1 correspondence
6. No, the members are countable

S1. 3
S2. -21
1. 14
2. -18
3. -14
4. -31
5. 24
6. 37
7. -168
8. -13
9. -34
10. -10

Problem Solving: -20°

Page 147

Review Exercises
1. -7
2. 13
3. -39
4. 0
5. A set is a well defined collection of objects
6. A set whose number of members is countable.

S1. -4
S2. -7
1. -2
2. -7
3. -8
4. 14
5. -2
6. 10
7. -22
8. -13
9. -25
10. -84

Problem Solving: $55

Page 148

Review Exercises
1. Ø
2. {0,1,4,5,8,9,10,12,15}
3. {10,15}
4. {1,4,8,9,10,11,12,15}
5. {1,4,8,9,12}
6. -49

S1. -14
S2. -3
1. 12
2. -3
3. 9
4. -28
5. 46
6. -48
7. -10
8. -11
9. -103
10. 15

Problem Solving: 27 ft.

Page 149

Review Exercises
1. -56
2. 22
3. -35
4. -9
5. -11
6. 4

S1. 48
S2. -126
1. 68
2. -64
3. 288
4. -368
5. -736
6. -24
7. -32
8. 72
9. 72
10. 330

Problem Solving: Floor 31

Page 150

Review Exercises
1. -11
2. -56
3. -2
4. 2
5. -35
6. 36

S1. -4
S2. 6
1. -16
2. 48
3. 15
4. -26
5. -3
6. -2
7. 1
8. -1
9. 3
10. 2

Problem Solving: -11°

Page 151

Review Exercises
1. -2
2. 2
3. -16
4. -1
5. -19
6. -2
7. 13
8. -12
9. -27
10. -1
11. -48
12. 76
13. 56
14. 48
15. -9
16. 42
17. 16
18. 3
19. -2
20. -20

Pre-Algebra—Solutions

Page 152

Review Exercises
1. {1,4}
2. {1,2,3,4,6,7}
3. {1,6}
4. {1,2,3,4,5,6,9}
5. {1,3,4,5,6,7,9}
6. {1,3,4,5,9}

S1. 3/10
S2. -1/10
1. -1/4
2. -1 1/6
3. -2
4. 3/8
5. -7/12
6. 2 1/2
7. 1 1/12
8. 1/8
9. -10
10. 7/15

Problem Solving: 50 sixth graders

Page 153

Review Exercises
1. -59
2. -36
3. -64
4. -28
5. 4
6. 28

S1. -.91
S2. -10.3
1. -12.81
2. 3.17
3. -4.284
4. 12.13
5. 2.37
6. .18
7. .426
8. -6.93
9. 2.01
10. -17.04

Problem Solving: 118 pounds

Page 154

Review Exercises
1. .4
2. -5.9
3. -7.8
4. -5/6
5. -1/10
6. 3/5

S1. 16
S2. -27
1. 216
2. 1
3. 16
4. 32
5. 7
6. 512
7. -1
8. 3,125
9. -125
10. 81

Problem Solving: 2

Page 155

Review Exercises
1. 49
2. 729
3. 36
4. 4
5. 1
6. 9

S1. 12^3
S2. 3^3
1. 2^6
2. $(-9)^3$
3. 16^4
4. 7^2 or $(-7)^2$
5. 10^2 or $(-10)^2$
6. 11^2 or $(-11)^2$
7. $(-1)^4$
8. 2^5
9. 2^4 or 4^2 or $(-2)^4$ or $(-4)^2$
10. 9^6

Problem Solving: (-3)

Page 156

Review Exercises
1. -9
2. -3
3. -7/12
4. 3.6
5. -1.04
6. -3/8

S1. 5
S2. 12
1. 4
2. 11
3. 1
4. 30
5. 10
6. 20
7. 13
8. 3
9. 16
10. 40

Problem Solving: -41

Page 157

Review Exercises
1. 36
2. -8
3. 6^4
4. 8
5. 13
6. 11

S1. 36
S2. 8
1. 2
2. 1
3. 25
4. 88
5. 144
6. 64
7. 54
8. 4
9. 144
10. 9

Problem Solving: 81

Page 158

Review Exercises
1. 13^4
2. 2^7
3. 2^6 or 8^2 or $(-2)^6$ or $(-8)^2$
4. $(-2)^4$
5. 2^3
6. 10^2 or $(-10)^2$
7. 4
8. 8
9. 5
10. 20
11. 3
12. 18
13. 22
14. 4
15. 85
16. 400
17. 63
18. 675
19. 25
20. 10

Page 159

Review Exercises
1. 49
2. 6
3. -2
4. -56
5. -1
6. 24

S1. 28
S2. 38
1. 7
2. 15
3. 34
4. 25
5. 12
6. 27
7. 14
8. 4
9. 28
10. 36

Problem Solving: -1 yard

Pre-Algebra—Solutions

Page 160

Review Exercises
1. {2,4}
2. {8}
3. Ø
4. {1,2,4,5,7,8,9}
5. {1,2,4,5,6,8,10}
6. {1,2,4,5,8}

S1. 4
S2. 9
1. 29
2. 6
3. 2
4. 32
5. 1
6. 7
7. 6
8. 12
9. 7
10. 24

Problem Solving: $29

Page 161

Review Exercises
1. 27
2. 28
3. -13
4. A well-defined collection of objects
5. -1 1/4
6. -.41

S1. commutative (addition)
S2. distributive
1. inverse property (addition)
2. associative (multiplication)
3. identity (addition)
4. inverse (multiplication)
5. commutative (addition)
6. commutative (multiplication)
7. associative (addition)
8. identity (multiplication)
9. distributive
10. inverse (addition)

Problem Solving: 19

Page 162

Review Exercises
1. 11
2. 17
3. 10
4. 23
5. 60
6. 42

S1. $3 + (7 + 9)$
S2. 15×7
1. 1
2. $3 \times 6 + 3 \times 2$
3. $12 + 9$
4. $(3 \times 9) \times 5$
5. $3(5 + 7)$
6. -9
7. 1
8. 5
9. $(3 + 5) + 6$
10. $3 \times 4 + 3 \times (-2)$

Problem Solving: $42

Page 163

Review Exercises
1. 10
2. 125
3. 21
4. -90
5. 5
6. -5/6

S1. 2.3×10^9
S2. 1.49×10^{-7}
1. 6.53×10^{11}
2. 1.597×10^5
3. 1.06×10^8
4. 7.216×10^{-6}
5. 1.096×10^9
6. 1.963×10^{-3}
7. 1.6×10^{-10}
8. 8×10^{-10}
9. 7×10^{12}
10. 1.287×10^{-7}

Problem Solving: 1.86×10^5 miles per second

Page 164

Review Exercises
1. 1.23×10^5
2. 3.21×10^{-4}
3. distributive
4. -17
5. -3
6. 35

S1. 7,032,000
S2. .000056
1. 230,000
2. .0000000913
3. .000012362
4. 517,000,000,000
5. 1,127
6. .003012
7. 6,670,000
8. 21,000
9. .00000007
10. 8,000,000

Problem Solving: 93,000,000 miles

Page 165

Review Exercises
1. 1.23×10^5
2. 5.6×10^{-6}
3. 2,760,000
4. .0000375
5. answers vary
6. answers vary

S1. 5/3
S2. 9/2
1. 7/2
2. 6/5
3. 6/5
4. 5/1
5. 6/5
6. 4/3
7. 7/3
8. 3/2
9. 1/2
10. 3/1

Problem Solving: 12/5

Page 166

Review Exercises
1. 2.7×10^{-4}
2. 2.916×10^6
3. 721,000
4. .0000623
5. 30
6. -1.32

S1. yes
S2. no
1. yes
2. no
3. yes
4. yes
5. yes
6. no
7. yes
8. no
9. yes
10. no

Problem Solving: 2

Page 167

Review Exercises
1. 5/3
2. yes ($4 \times 10 = 8 \times 5$)
3. no ($5 \times 5 \neq 2 \times 7$)
4. 16
5. 48
6. 275

S1. 1
S2. 16
1. 3
2. 2
3. 42
4. 6 2/5
5. 6
6. 21
7. 18
8. 2 4/5
9. 1.2
10. 2 1/3

Problem Solving: -32°

Page 168

Review Exercises
1. yes ($4 \times 9 = 3 \times 12$)
2. n = 30
3. n = 12
4. 2.34×10^8
5. $2.35 \times 10\text{-}3$
6. 720,000

S1. 204 miles
S2. $12
1. 2 gallons
2. $17.50
3. 15 girls
4. 40 miles
5. 2 4/5 pounds

Problem Solving: 12°

Page 169

Review Exercises
1. 3/1
2. 12/5
3. 8/3
4. yes, $15 \times 24 = 12 \times 30$
5. no, $7 \times 9 \neq 8 \times 8$
6. yes, $5 \times 9 = 3 \times 15$
7. 4
8. 33
9. 5
10. 35
11. 4 1/2
12. 9
13. 20
14. 3
15. 9
16. $8.40
17. 30 boys
18. 70 miles
19. 2 gallons
20. 4 pounds

Page 170

Review Exercises
1. 21
2. 3 3/5
3. 9
4. 18
5. 26
6. -33

S1. .2, 1/5
S2. .09, 9/100
1. .16, 4/25
2. .06, 3/50
3. .75, 3/4
4. .4, 2/5
5. .01, 1/100
6. .45, 9/20
7. .12, 3/25
8. .05, 1/20
9. .5, 1/2
10. .13, 13/100

Problem Solving: 19/20

Page 171

Review Exercises
1. .8
2. .07
3. 1/4
4. 109.2
5. 128
6. 18

S1. 17.5
S2. 150
1. 4.32
2. 51
3. 15
4. 112.5
5. 32
6. 80
7. 10
8. 216
9. 112.5
10. 13.2

Problem Solving: 34 correct

Page 172

Review Exercises
1. 46.5
2. 24
3. .75
4. 70%
5. 135
6. 600

S1. 25%
S2. 75%
1. 25%
2. 80%
3. 50%
4. 90%
5. 60%
6. 75%
7. 75%
8. 75%
9. 80%
10. 95%

Problem Solving: 75%

Page 173

Review Exercises
1. 3.2
2. 32
3. 75%
4. 90%
5. 50.04
6. 200

S1. 20
S2. 30
1. 48
2. 80
3. 25
4. 4
5. 15
6. 20
7. 60
8. 75
9. 45
10. 125

Problem Solving: 25 students

Page 174

Review Exercises
1. .00072
2. 2 19/10,000
3. 60%
4. 15
5. .021
6. 80

S1. 20 questions
S2. 75%
1. $25
2. $1,600
3. 30
4. 90%
5. 250 cows
6. 25
7. $240
8. 180 boys
9. 20%
10. $210

Problem Solving: $57,600

Page 175

Review Exercises
1. 32.744
2. 2.358
3. 21.98
4. .01248
5. .48
6. 8.1

S1. 30
S2. 30
1. 20%
2. 90%
3. 30 students
4. 18 passes
5. 20% are red
6. 75%
7. 250 students
8. $25
9. 60%
10. $691.20

Problem Solving: 97

Page 176

Review Exercises
1. 13%
2. 3%
3. 70%
4. 19%
5. 60%
6. .08, 2/25
7. .18, 9/50
8. .8, 4/5
9. 2.22
10. 128
11. 80%
12. 75%
13. 12
14. 75
15. 80%
16. 60%
17. 256 girls
18. 26 games
19. 75%
20. 150 students

Page 177

Review Exercises
1. 70%
2. 80%
3. 3/25
4. 12
5. 25%
6. 25

S1. 1, 30, 2, 15, 3, 10, 5, 6
S2. 1, 36, 2, 18, 3, 12, 4, 9, 6
1. 1, 100, 2, 50, 4, 25, 5, 20, 10
2. 1, 42, 2, 21, 3, 14, 6,7
3. 1, 70, 2, 35, 5, 14, 7, 10
4. 1, 81, 3, 27, 9
5. 1, 50, 2, 25, 5, 10
6. 1, 40, 2, 20, 4, 10, 5, 8
7. 1, 75, 3, 25, 5, 15
8. 1, 90, 2, 45, 3, 30, 5, 18, 6, 15, 9, 10
9. 1, 20, 2, 10, 4, 5
10. 1, 50, 2, 25, 5, 10

Problem Solving: 54 correct

Page 178

Review Exercises
1. -18
2. 24
3. 20
4. 7 1/2
5. 15
6. 25%

S1. 2
S2. 4
1. 2
2. 3
3. 14
4. 16
5. 20
6. 10
7. 5
8. 12
9. 12
10. 20

Problem Solving: 5,879,000,000,000 miles

Page 179

Review Exercises
1. 1.2×10^{-6}
2. 4.96×10^{8}
3. 13,200,000
4. .00000464
5. 1,60,2,30,3,10,5,6
6. 4

S1. 4, 6, 8, 10
S2. 0, 12, 18, 30
1. 10, 15, 20, 25
2. 0, 6, 12, 15
3. 0, 30, 40, 50
4. 0, 4, 8
5. 22, 44
6. 24, 32, 40
7. 60, 80, 100
8. 14, 28, 35
9. 90, 120, 150
10. 27, 45

Problem Solving: 120 pitches

Page 180

Review Exercises
1. 1, 30, 2, 15, 3, 10, 5, 6
2. 4
3. 0, 8, 16, 24, 32, 40
4. 3
5. 25%
6. 35

S1. 12
S2. 24
1. 15
2. 30
3. 60
4. 30
5. 36
6. 60
7. 48
8. 40
9. 36
10. 60

Problem Solving: $12.96

Page 181

1. 1, 24, 2, 12, 3, 8, 4, 6
2. 1, 16, 2, 8, 4
3. 1, 32, 2, 16, 4, 8
4. 1, 28, 2, 14, 4, 7
5. 1, 70, 2, 35, 7, 10
6. 1, 25, 5
7. 4
8. 12
9. 20
10. 5
11. 7
12. 18
13. 9, 12, 15
14. 9, 27, 36, 45
15. 15, 30, 45, 60
16. 12
17. 60
18. 60
19. 12
20. 24

Page 182

Review Exercises
1. 8
2. 12
3. 1, 28, 2, 14, 4, 7
4. 0, 12, 24, 36, 48, 60
5. yes, $3 \times 8 = 4 \times 6$
6. 9

S1. 0
S2. -7, -2, 10
1. 5, 9
2. 3, 4
3. 2, 4, 7
4. -5, -8
5. 10, 9, -6, 6
6. 9, -7, 1
7. -8, 8, 0, -3
8. 0, 7, 8
9. -6, 4, -3
10. 5, -3, 9, -8

Problem Solving: 8°

Page 183

Review Exercises
1. {4,6,8,10}
2. {1,3,4,5,6,8,9,10}
3. {2,4,5,6,8,9,10}
4. {4,5,6,8,10}
5. no, cannot be paired in 1-1 correspondence
6. no, they have members in common

S1. 1
S2. 7
1. 3
2. 5
3. 5
4. 18
5. 4
6. 11
7. 8
8. 7
9. 4
10. 12

Problem Solving: 512 miles

Page 184

Review Exercises
1. 7
2. -22
3. 8
4. -42
5. 5
6. -2

S1. (2, 1)
S2. (-4, 2)
1. (6, 3)
2. (2, -5)
3. (-7, -3)
4. (-5, 1)
5. (4, 6)
6. (4, -3)
7. (-3, -5)
8. (-2, 2)
9. (2, 1)
10. (-6, 7)

Problem Solving: $24

Page 185

Review Exercises
1. -7/15
2. -.68
3. 3/8
4. 1
5. -5
6. 2

S1. B
S2. A
1. C
2. D
3. F
4. M
5. E
6. J
7. H
8. G
9. K
10. I

Problem Solving: 95%

Page 186

Review Exercises
1. 32
2. 22
3. 5
4. 72
5. 1.7×10^{-4}
6. 2.13×10^{5}

S1. 1/3
S2. 3/2
1. -3/2
2. 1/3
3. -2
4. 5/8
5. 1/2
6. 7/2
7. 6/5
8. 4/5
9. 3/2
10. -3/2

Problem Solving: 320 girls

Page 187

1. -4
2. 9, 0, 10
3. 3, 2, -7
4. 8, 1, -4, -8
5. 3
6. 6
7. 5
8. 8
9. (5, 3)
10. (-6, 1)
11. (4, -4)
12. (-6, -5)
13. (2, 6)
14. 5/3
15. B
16. I
17. H
18. D
19. A
20. 1/8

Page 188

Review Exercises
1. 6
2. -18
3. -24
4. -16
5. -7
6. -8

Problem Solving: $6.00

S1.

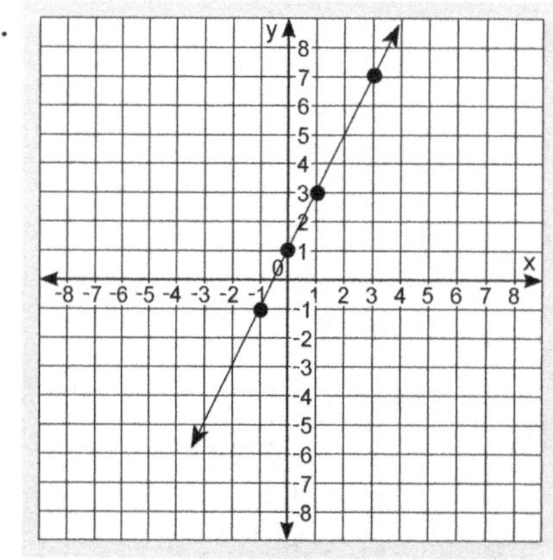

S2.

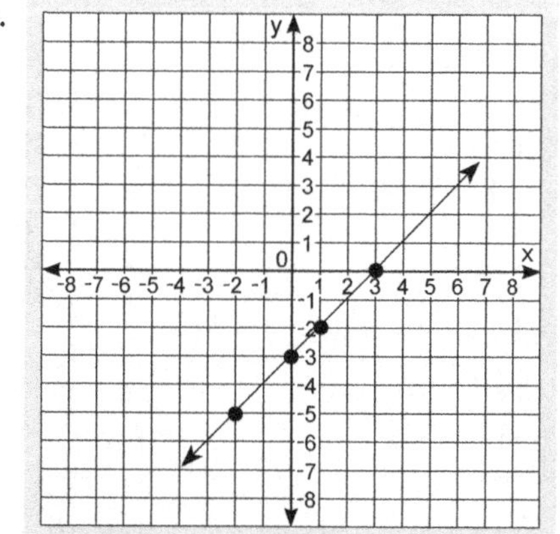

1.

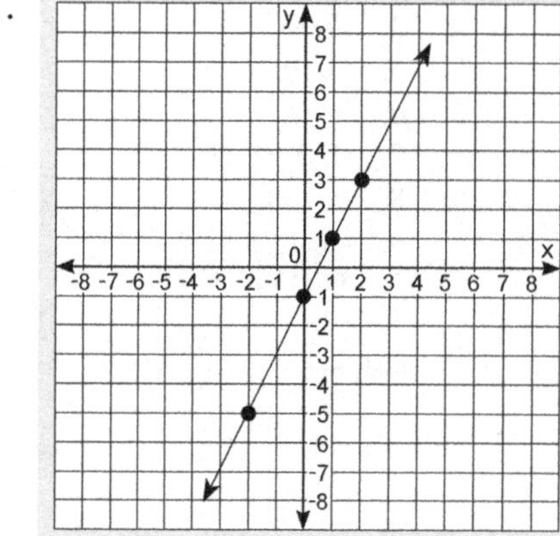

2.

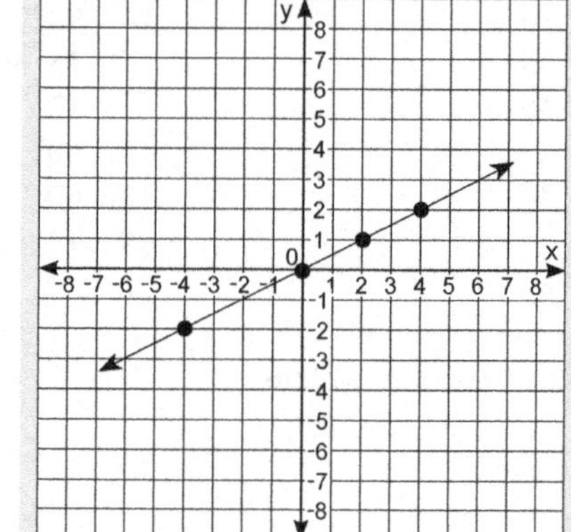

Page 189

Review Exercises
1. 8 2/5
2. 3
3. 75
4. -1/2
5. -3
6. -2.1

Problem Solving: 80%

S1.

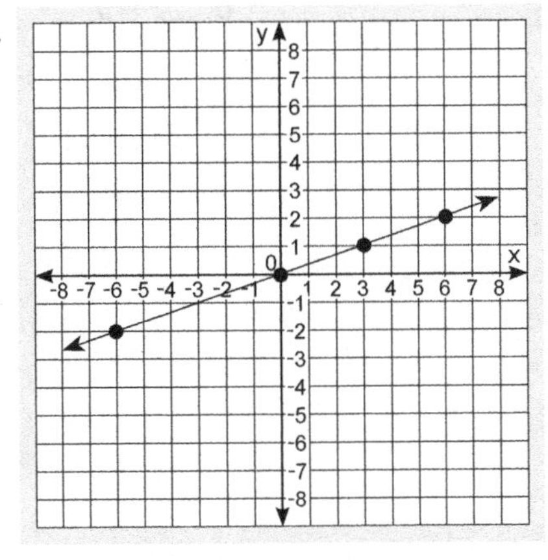

S2.

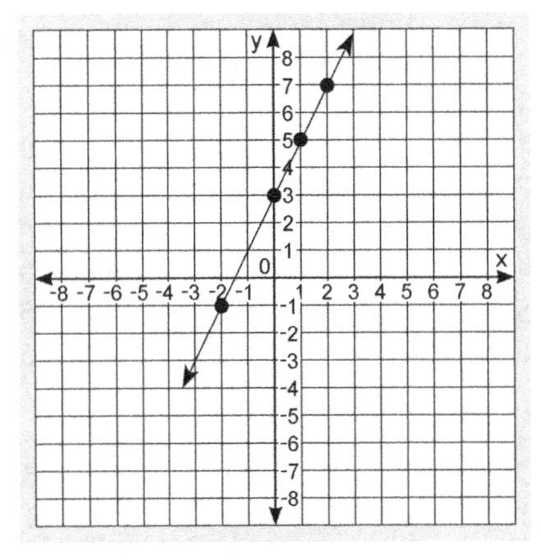

1.

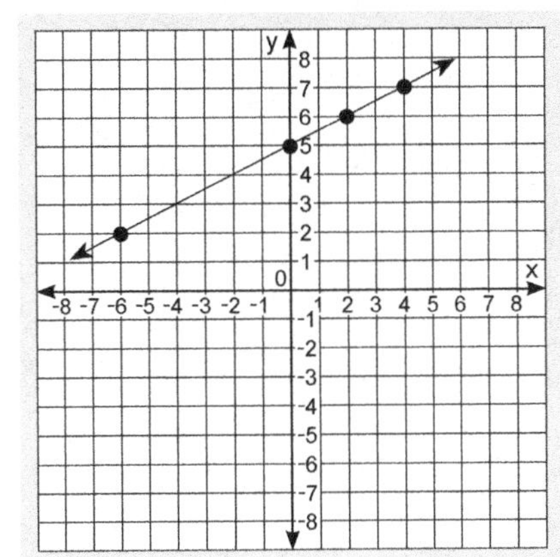

2.

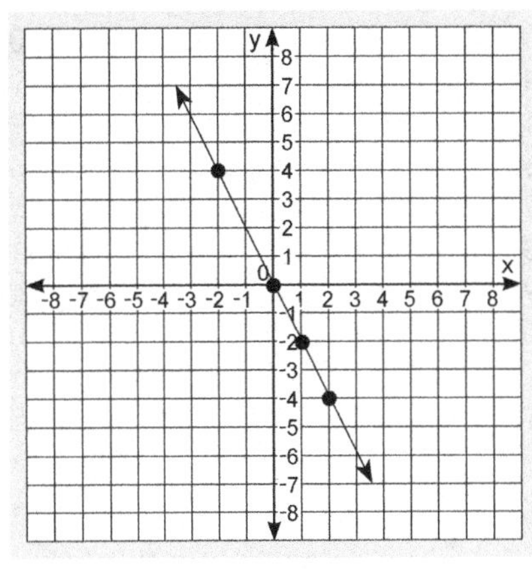

Page 190

Review Exercises
1. 1.72×10^6
2. 3.8×10^{-7}
3. 196,300,000
4. .00034
5. -10
6. .19

S1. 5
S2. 32
1. -3
2. 40
3. -13
4. -5
5. 13
6. -12
7. 34
8. 4
9. -11
10. -27

Problem Solving: 7/3

Page 191

Review Exercises
1. 10
2. 23
3. 102
4. -8
5. 6
6. 21

S1. 12
S2. 3
1. -14
2. -12
3. -9
4. -5
5. 18
6. -6
7. 2
8. 20
9. -15
10. 14

Problem Solving: 9/4

Page 192

Review Exercises
1. 1
2. 72
3. 2
4. 2.1×10^5
5. 3.16×10^{-3}
6. 16

S1. 7
S2. 4
1. -1
2. -4
3. -15
4. 75
5. 12
6. 1
7. 2
8. 50
9. 10
10. -3

Problem Solving: 200 students

Page 193

Review Exercises
1. 1/10
2. -4/5
3. -1 1/3
4. -3
5. -.64
6. -1.1

S1. 3
S2. 45
1. 8
2. -3
3. 10
4. -3
5. 5
6. -4
7. -10
8. -7
9. -3
10. 10

Problem Solving: 14

Pre-Algebra—Solutions

Page 194

Review Exercises
1. 216
2. 1
3. 9
4. 13
5. 12
6. 158

S1. -2
S2. 3
1. 7
2. 2
3. 4
4. 4
5. 3
6. 1
7. -1
8. 4
9. 2
10. 6

Problem Solving: 20%

Page 195

Review Exercises
1. 20
2. no, $4 \times 5 \neq 3 \times 7$
3. 8/3
4. 60
5. 25%
6. 35

S1. 4
S2. 15
1. -2
2. -3
3. 10
4. -5
5. 12
6. 1
7. 7
8. 3
9. -25
10. -10

Problem Solving:
8 blue marbles

Page 196

Review Exercises
1. 1, 48, 2, 24, 3, 16, 4, 12, 6, 8
2. 8
3. 30
4. 5
5. 20
6. -5

S1. $2x - 7 = 12$
S2. $3x + 2 = 30$
1. $2x + 5 = 14$
2. $4x - 6 = 10$
3. $4x - 5 = 12$
4. $x/3 - 4 = 2x + 8$
5. $2(x + 2) = 10$
6. $5x - 3 = 17$
7. $2x - 6 = 15$
8. $3x - 2 = 2x + 7$
9. $x + 4 = 7 + -12$
10. $n/5 = 25$

Problem Solving: 227.5 miles

Page 197

Review Exercises
1. 7
2. -4
3. 12
4. 5
5. 15
6. 3

S1. 11
S2. 12
1. 36
2. -3
3. 7
4. 4
5. 3

Problem Solving: $12,500

Page 198

Review Exercises
1. .000000361
2. 1.27×10^{-6}
3. 7.29×10^{8}

S1. Kevin is 12
Amir is 18
S2. Short piece = 11 inches
Long piece = 33 inches
1. Bob earned $20
Bill earned $46
2. Monday, $91
Tuesday, $121
3. 12 years old
4. Weekly salary is $140
5. John is 30

Problem Solving: 89

Pag199

Review Exercises
1. -11
2. 20
3. 44
4. -9
5. 3
6. 11

S1. 6
S2. Bart is 110 pounds
Bob is 160 pounds
1. -9
2. 6
3. -24
4. Ellen is 11
Roy is 33
5. 3

Problem Solving:
600 miles per hour

Page 200

1. -7
2. 13
3. 56
4. 3
5. -5
6. -18
7. -7
8. 15
9. 8
10. 8
11. -11
12. 21
13. 12
14. -5
15. 6
16. 6
17. Sue, $22
Ann, $44
18. Ron is 28
Bill is 36
19. -2
20. 18

Page 201

Review Exercises
1. 0,8,16,24,32,40,48
2. 1, 60, 2, 30, 3, 20, 4, 15, 5, 12, 6, 10
3. 20
4. 6×10^{-6}
5. 2.1×10^{6}
6. .0021

S1. 3/12 = 1/4
S2. 7/12
1. 2/12 = 1/6
2. 1/12
3. 9/12 = 3/4
4. 10/12 = 5/6
5. 9/12 = 3/4
6. 5/12
7. 10/12 = 5/6
8. 6/12 = 1/2
9. 6/12 = 1/2
10. 4/12 = 1/3

Problem Solving: -3

Pre-Algebra—Solutions

Page 202

Review Exercises
1. -10
2. -25
3. -10
4. 4
5. 15
6. 9

S1. 1/8
S2. 4/8 = 1/2
1. 1/8
2. 7/8
3. 4/8 = 1/2
4. 4/8 = 1/2
5. 2/8 = 1/4
6. 0/8
7. 2/8 = 1/4
8. 5/8
9. 5/8
10. 4/8 = 1/2

Problem Solving: 20 pounds

Page 203

Review Exercises
1. 30%
2. 3%
3. 60%
4. 2
5. 25%
6. 20

S1. range 7, mode 4
S2. range 6, mode 6
1. range 6, mode 7
2. range 14, mode 30
3. range 10, mode 3
4. range 19, mode 9
5. range 7, mode 3
6. range 11, mode 8
7. range 9, mode 2
8. range 6, mode 91
9. range 9, mode 2
10. range 19, mode 2

Problem Solving: $.38

Page 204

Review Exercises
1. 8/5
2. no, $7 \times 11 \neq 8 \times 9$
3. 20
4. 1.28×10^6
5. 9.62×10^{-5}
6. .000062

S1. mean 3, median 3
S2. mean 4, median 4
1. mean 3, median 2
2. mean 3, median 2
3. mean 15, median 15
4. mean 2, median 1
5. mean 6, median 6
6. mean 124, median 126
7. mean 4, median 4
8. mean 4, median 4
9. mean 5, median 4
10. mean 50, median 50

Problem Solving: 32 students

Page 205

Review Exercises
1. 21
2. 14
3. 15
4. 25
5. 84
6. 11

S1. 6
S2. 2
1. 5
2. 6
3. 3
4. 2
5. 8
6. 4
7. 10
8. 2
9. 5
10. 5

Problem Solving:
186,000 miles per second

Page 206

1. 3/10
2. 4/10 = 2/5
3. 5/10 = 1/2
4. 7/10
5. 8/10 = 4/5
6. 6/10 = 3/5
7. 1/8
8. 4/8 = 1/2
9. 5/8
10. 2/8 = 1/4
11. 4/8 = 1/2
12. 2/8 = 1/4
13. 8
14. 4
15. 5
16. 4
17. 2
18. 4
19. 7
20. 3

Page 207

1. {2,3,4}
2. {0,1,2,3,4,5,6,8,9}
3. {1,2,4,5}
4. 3
5. -23
6. 36
7. 8
8. -.55
9. -9/10
10. 125
11. 7
12. 33
13. 22
14. 68
15. 2
16. 20
17. commutative (addition)
18. distributive
19. 1.28×10^9
20. 6.53×10^{-6}

Page 208

21. 60,900,000
22. .00000762
23. 9/5
24. no, $8 \times 3 \neq 4 \times 7$
25. 21
26. 6 blue marbles
27. 4
28. 25%
29. 32
30. 24
31. 40%
32. 30
33. 1, 40, 2, 20, 4, 10, 5, 8
34. 20
35. 0, 6, 12, 18, 24, 30, 36
36. 24
37. -8, -2, 0
38. -5, 9, -1, -7
39. -6, 4, -3
40. 9, 8, 1

Page 209

41. (6, -3)
42. (-5, 1)
43. (4, -3)
44. (-4, -2)
45. (4, 6)
46. H
47. G
48. L
49. J
50. K
51. 2/3
52. 3/8

53.

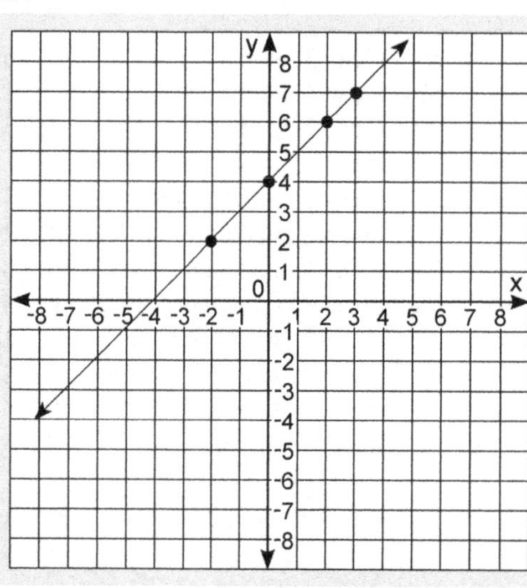

54.

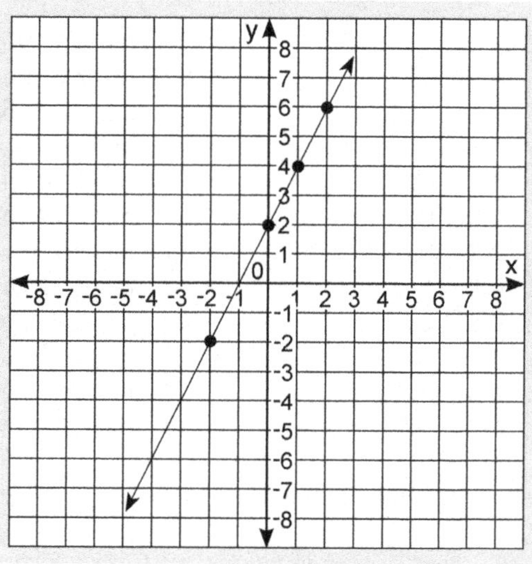
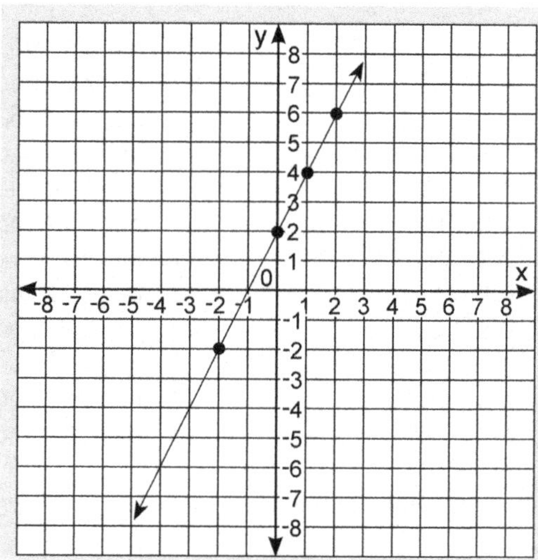

Page 210

55. 9
56. -15
57. 18
58. -3
59. 6
60. -3
61. 24
62. 4
63. -6
64. 1
65. 9
66. 12
67. 50
68. Jane, $16
    Sue, $48
69. Maria is 22
    Al is 29
70. 9

Page 211

71. 5/16
72. 6/16 = 3/8
73. 7/16
74. 11/16
75. 1/8
76. 4/8 = 1/2
77. 4/8 = 1/2
78. 3/8
79. 8
80. 3
81. 5
82. 3
83. 4
84. 8
85. 14
86. 6

Page 212

87. 1 1/4
88. 3 1/2
89. 5 7/10
90. 2
91. 8 3/4
92. 4 2/3
93. 3 2/3
94. 19.42
95. 5.025
96. 70.32
97. 15.33
98. .432
99. 1.34
100. .03

A

absolute value The distance of a number from 0 on the number line.
The absolute value is always positive.

acute angle An angle with a measure of less than 90 degrees.

adjacent Next to.

algebraic expression A mathematical expression that contains at least one variable.

angle Any two rays that share an endpoint will form an angle.

associative properties For any a, b, c:
addition: $(a + b) + c = a + (b + c)$
multiplication: $(ab)c = a(bc)$

B

base The number being multiplied. In an expression such as 4^2, 4 is the base.

C

coefficient A number that multiplies the variable. In the term 7x, 7 is the coefficient of x.

commutative properties For any a, b:
addition: $a + b = b + a$
multiplication: $ab = ba$

complementary angles Two angles that have measures whose sum is 90 degrees.

congruent Two figures having exactly the same size and shape.

coordinate plane The plane which contains the x- and y-axes. It is divided into 4 quadrants.
Also called coordinate system and coordinate grid.

coordinates An ordered pair of numbers that identify a point on a coordinate plane.

D

data Information that is organized for analysis.

degree A unit that is used in measuring angles.

denominator The bottom number of a fraction that tells the number of equal parts into which a whole is divided.

disjoint sets Sets that have no members in common. {1,2,3} and {4,5,6} are disjoint sets.

distributive property For real numbers a, b, and c: $a(b + c) = ab + ac$.

E

element of a set Member of a set.

empty set The set that has no members. Also called the null set and written Ø or { }.

equation A mathematical sentence that contains an equal sign (=) and states that one expression is equal to another expression.

equivalent Having the same value.

exponent A number that indicates the number of times a given base is used as a factor. In the expression n^2, 2 is the exponent.

expression Variables, numbers, and symbols that show a mathematical relationship.

extremes of a proportion In the proportion $\frac{a}{b} = \frac{c}{d}$, a and d are the extremes.

F

factor An integer that divides evenly into another.

finite Something that is countable.

formula A general mathematical statement or rule. Used often in algebra and geometry.

function A set of ordered pairs that pairs each x-value with one and only one y-value. (0,2), (-1,6), (4,-2), (-3,4) is a function.

G

graph To show points named by numbers or ordered pairs on a number line or coordinate plane. Also, a drawing to show the relationship between sets of data.

greatest common factor The largest common factor of two or more numbers. Also written GCF. The greatest common factor of 15 and 25 is 5.

grouping symbols Symbols that indicate the order in which mathematical operations should take place. Examples include parentheses ( ), brackets [ ], braces { }, and fraction bars —.

H

hypotenuse The side opposite the right angle in a right triangle.

I

identity properties of addition and multiplication For any real number a:
addition: $a + 0 = 0 + a = a$
multiplication: $1 \times a = a \times 1 = a$

inequality A mathematical sentence that states one expression is greater than or less than another. Inequality symbols are read as follows:
$<$ less than
$\leq$ less than or equal to
$>$ greater than
$\geq$ greater than or equal to

infinite Having no boundaries or limits. Uncountable.

integers Numbers in a set. ...-3, -2, -1, 0, 1, 2, 3...

intersection of sets If A and B are sets, then A intersection B is the set whose members are included in both sets A and B, and is written $A \cap B$. If set A = {1,2,3,4} and set B = {1,3,5}, then $A \cap B = \{1,3\}$

inverse properties of addition and multiplication For any number a:
addition: $a + -a = 0$
multiplication: $a \times 1/a = 1$ ($a \neq 0$)

inverse operations Operations that "undo" each other. Addition and subtraction are inverse operations, and multiplication and division are inverse operations.

L

least common multiple The least common multiple of two or more whole numbers is the smallest whole number, other than zero, that they all divide into evenly. Also written as LCM. The least common multiple of 12 and 8 is 24.

linear equation An equation whose graph is a straight line.

M

mean In statistics, the sum of a set of numbers divided by the number of elements in the set. Sometimes referred to as average.

means of a proportion In the proportion $\frac{a}{b} = \frac{c}{d}$, b and c are the means.

median In statistics, the middle number of a set of numbers when the numbers are arranged in order of least to greatest. If there are two middle numbers, find their mean.

mode In statistics, the number that appears most frequently. Sometimes there is no mode. There may also be more than one mode.

multiple The product of a whole number and another whole number.

N

natural numbers Numbers in the set 1, 2, 3, 4,... Also called counting numbers.

negative numbers Numbers that are less than zero.

null set The set that has no members. Also called the empty set and written Ø or { }.

number line A line that represents numbers as points.

numerator The top part of a fraction.

O

obtuse angle An angle whose measure is greater than 90° and less than 180°.

opposites Numbers that are the same distance from zero, but are on opposite sides of zero on a number line. 4 and -4 are opposites.

order of operations The order of steps to be used when simplifying expressions.

1. Evaluate within grouping symbols.
2. Eliminate all exponents.
3. Multiply and divide in order from left to right.
4. Add and subtract in order from left to right.

ordered pair A pair of numbers (x,y) that represent a point on the coordinate plane. The first number is the x-coordinate and the second number is the y-coordinate.

origin The point where the x-axis and the y-axis intersect in a coordinate plane. Written as (0,0).

outcome One of the possible events in a probability situation.

P

parallel lines Lines in a plane that do not intersect. They stay the same distance apart.

percent Hundredths or per hundred. Written %.

perimeter The distance around a figure.

perpendicular lines Lines in the same plane that intersect at a right (90°) angle.

pi The ratio of the circumference of a circle to its diameter. Written π. The approximate value for π is 3.14 as a decimal and $\frac{22}{7}$ as a fraction.

plane A flat surface that extends infinitely in all directions.

point An exact position in space. Points also represent numbers on a number line or coordinate plane.

positive number Any number that is greater than 0.

power An exponent.

prime number A whole number greater than 1 whose only factors are 1 and itself.

probability What chance, or how likely it is for an event to occur. It is the ratio of the ways a certain outcome can occur and the number of possible outcomes.

proportion An equation that states that two ratios are equal. $\frac{4}{8} = \frac{2}{4}$ is a proportion.

Pythagorean theorem In a right triangle, if c is the hypotenuse, and a and b are the other two legs, then $a^2 + b^2 = c^2$.

Q

quadrant One of the four regions into which the x-axis and y-axis divide a coordinate plane.

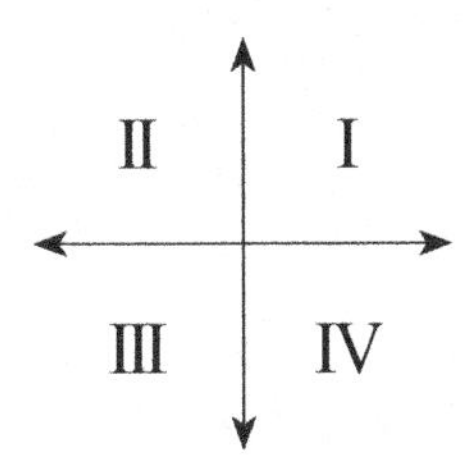

R

range The difference between the greatest number and the least number in a set of numbers.

ratio A comparison of two numbers using division. Written a:b, a to b, and a/b.

reciprocals Two numbers whose product is 1. $\frac{2}{3}$ and $\frac{3}{2}$ are reciprocals because $\frac{2}{3} \times \frac{3}{2} = 1$.

reduce To express a fraction in its lowest terms.

relation Any set of ordered pairs.

right angle An angle that has a measure of 90°.

rise The change in y going from one point to another on a coordinate plane. The vertical change.

run The change in x going from one point to another on a coordinate plane. The horizontal change.

S

scientific notation A number written as the product of a numbers between 1 and 10 and a power of ten. In scientific notation, $7{,}000 = 7 \times 10^3$.

set A well-defined collection of objects.

slope Refers to the slant of a line. It is the ratio of rise to run.

solution A number that can be substituted for a variable to make an equation true.

square root Written $\sqrt{}$. The $\sqrt{36} = 6$ because $6 \times 6 = 36$.

statistics Involves data that is gathered about people or things and is used for analysis.

subset If all the members of set A are members of set B, then set A is a subset of set B. Written $A \subset B$. If set A = {1,2,3} and set B = {0,1,2,3,5,8}, set A is a subset of set B because all of the members of a set A are also members of set B.

U

union of sets If A and B are sets, the union of set A and set B is the set whose members are included in set A, or set B, or both set A and set B. A union B is written $A \cup B$. If set = {1,2,3,4} and set B = {1,3,5,7}, then $A \cup B$ = {1,2,3,4,5,7}.

universal set The set which contains all the other sets which are under consideration.

variable A letter that represents a number.

Venn diagram A type of diagram that shows how certain sets are related.

vertex The point at which two lines, line segments, or rays meet to form an angle.

W

whole number Any number in the set 0, 1, 2, 3, 4...

X

x-axis The horizontal axis on a coordinate plane.

x-coordinate The first number in an ordered pair. Also called the abscissa.

Y

y-axis The vertical axis on a coordinate plane.

y-coordinate The second number in an ordered pair. Also called the ordinate.

Important Symbols

- $<$ less than
- $\leq$ less than or equal to
- $>$ greater than
- $\geq$ greater than or equal to
- $=$ equal to
- $\neq$ not equal to
- $\cong$ congruent to
- ( ) parenthesis
- [ ] brackets
- { } braces
- ... and so on
- • or × multiply
- ∞ infinity
- a^n the nth power of a number
- $\sqrt{\ }$ square root
- Ø, { } the empty set or null set
- $\therefore$ therefore
- $^\circ$ degree
- π pi
- { } set
- | | absolute value
- $.\overline{n}$ repeating decimal symbol
- 1/a the reciprocal of a number
- % percent
- (x,y) ordered pair
- $\perp$ perpendicular
- | | parallel to
- $\angle$ angle
- $\in$ element of
- $\notin$ not an element of
- $\cap$ intersection
- $\cup$ union
- $\subset$ subset of
- $\not\subset$ not a subset of
- $\triangle$ triangle

Multiplication Table

x	2	3	4	5	6	7	8	9	10	11	12
2	4	6	8	10	12	14	16	18	20	22	24
3	6	9	12	15	18	21	24	27	30	33	36
4	8	12	16	20	24	28	32	36	40	44	48
5	10	15	20	25	30	35	40	45	50	55	60
6	12	18	24	30	36	42	48	54	60	66	72
7	14	21	28	35	42	49	56	63	70	77	84
8	16	24	32	40	48	56	64	72	80	88	96
9	18	27	36	45	54	63	72	81	90	99	108
10	20	30	40	50	60	70	80	90	100	110	120
11	22	33	44	55	66	77	88	99	110	121	132
12	24	36	48	60	72	84	96	108	120	132	144

Commonly Used Prime Numbers

2	3	5	7	11	13	17	19	23	29
31	37	41	43	47	53	59	61	67	71
73	79	83	89	97	101	103	107	109	113
127	131	137	139	149	151	157	163	167	173
179	181	191	193	197	199	211	223	227	229
233	239	241	251	257	263	269	271	277	281
283	293	307	311	313	317	331	337	347	349
353	359	367	373	379	383	389	397	401	409
419	421	431	433	439	443	449	547	461	463
467	479	487	491	499	503	509	521	523	541
547	557	563	569	571	577	587	593	599	601
607	613	617	619	631	641	643	647	653	659
661	673	677	683	691	701	709	719	727	733
739	743	751	757	761	769	773	787	797	809
811	821	823	827	829	839	853	857	859	863
877	881	883	887	907	911	919	929	937	941
947	953	967	971	977	983	991	997	1009	1013

Squares and Square Roots

No.	Square	Square Root	No.	Square	Square Root	No.	Square	Square Root
1	1	1.000	51	2,601	7.141	101	10201	10.050
2	4	1.414	52	2,704	7.211	102	10,404	10.100
3	9	1.732	53	2,809	7.280	103	10,609	10.149
4	16	2.000	54	2,916	7.348	104	10,816	10.198
5	25	2.236	55	3,025	7.416	105	11,025	10.247
6	36	2.449	56	3,136	7.483	106	11,236	10.296
7	49	2.646	57	3,249	7.550	107	11,449	10.344
8	64	2.828	58	3,364	7.616	108	11,664	10.392
9	81	3.000	59	3,481	7.681	109	11,881	10.440
10	100	3.162	60	3,600	7.746	110	12,100	10.488
11	121	3.317	61	3,721	7.810	111	12,321	10.536
12	144	3.464	62	3,844	7.874	112	12,544	10.583
13	169	3.606	63	3,969	7.937	113	12,769	10.630
14	196	3.742	64	4,096	8.000	114	12,996	10.677
15	225	3.873	65	4,225	8.062	115	13,225	10.724
16	256	4.000	66	4,356	8.124	116	13,456	10.770
17	289	4.123	67	4,489	8.185	117	13,689	10.817
18	324	4.243	68	4,624	8.246	118	13,924	10.863
19	361	4.359	69	4,761	8.307	119	14,161	10.909
20	400	4.472	70	4,900	8.367	120	14,400	10.954
21	441	4.583	71	5,041	8.426	121	14,641	11.000
22	484	4.690	72	5,184	8.485	122	14,884	11.045
23	529	4.796	73	5,329	8.544	123	15,129	11.091
24	576	4.899	74	5,476	8.602	124	15,376	11.136
25	625	5.000	75	5,625	8.660	125	15,625	11.180
26	676	5.099	76	5,776	8.718	126	15,876	11.225
27	729	5.196	77	5,929	8.775	127	16,129	11.269
28	784	5.292	78	6,084	8.832	128	16,384	11.314
29	841	5.385	79	6,241	8.888	129	16,641	11.358
30	900	5.477	80	6,400	8.944	130	16,900	11.402
31	961	5.568	81	6,561	9.000	131	17,161	11.446
32	1,024	5.657	82	6,724	9.055	132	17,424	11.489
33	1,089	5.745	83	6,889	9.110	133	17,689	11.533
34	1,156	5.831	84	7,056	9.165	134	17,956	11.576
35	1,225	5.916	85	7,225	9.220	135	18,225	11.619
36	1,296	6.000	86	7,396	9.274	136	18,496	11.662
37	1,369	6.083	87	7,569	9.327	137	18,769	11.705
38	1,444	6.164	88	7,744	9.381	138	19,044	11.747
39	1,521	6.245	89	7,921	9.434	139	19,321	11.790
40	1,600	6.325	90	8,100	9.487	140	19,600	11.832
41	1,681	6.403	91	8,281	9.539	141	19,881	11.874
42	1,764	6.481	92	8,464	9.592	142	20,164	11.916
43	1,849	6.557	93	8,649	9.644	143	20,449	11.958
44	1,936	6.633	94	8,836	9.695	144	20,736	12.000
45	2,025	6.708	95	9,025	9.747	145	21,025	12.042
46	2,116	6.782	96	9,216	9.798	146	21,316	12.083
47	2,209	6.856	97	9,409	9.849	147	21,609	12.124
48	2,304	6.928	98	9,604	9.899	148	21,904	12.166
49	2,401	7.000	99	9,801	9.950	149	22,201	12.207
50	2,500	7.071	100	10,000	10.000	150	22,500	12.247

Fraction/Decimal Equivalents

Fraction	Decimal	Fraction	Decimal
$\frac{1}{2}$	0.5	$\frac{5}{10}$	0.5
$\frac{1}{3}$	0.3	$\frac{6}{10}$	0.6
$\frac{2}{3}$	0.6	$\frac{7}{10}$	0.7
$\frac{1}{4}$	0.25	$\frac{8}{10}$	0.8
$\frac{2}{4}$	0.5	$\frac{9}{10}$	0.9
$\frac{3}{4}$	0.75	$\frac{1}{16}$	0.0625
$\frac{1}{5}$	0.2	$\frac{2}{16}$	0.125
$\frac{2}{5}$	0.4	$\frac{3}{16}$	0.1875
$\frac{3}{5}$	0.6	$\frac{4}{16}$	0.25
$\frac{4}{5}$	0.8	$\frac{5}{16}$	0.3125
$\frac{1}{8}$	0.125	$\frac{6}{16}$	0.375
$\frac{2}{8}$	0.25	$\frac{7}{16}$	0.4375
$\frac{3}{8}$	0.375	$\frac{8}{16}$	0.5
$\frac{4}{8}$	0.5	$\frac{9}{16}$	0.5625
$\frac{5}{8}$	0.625	$\frac{10}{16}$	0.625
$\frac{6}{8}$	0.75	$\frac{11}{16}$	0.6875
$\frac{7}{8}$	0.875	$\frac{12}{16}$	0.75
$\frac{1}{10}$	0.1	$\frac{13}{16}$	0.8125
$\frac{2}{10}$	0.2	$\frac{14}{16}$	0.875
$\frac{3}{10}$	0.3	$\frac{15}{16}$	0.9375
$\frac{4}{10}$	0.4		

Made in the USA
Columbia, SC
16 August 2019